AS

Dan Evans and Alex Watts,
Abbey College, Manchester
Series Editor: Kevin Byrne

Chemistry

Where to find the information you need

SUCCESS OR YOUR MONEY BACK

Letts' market leading series AS in a Week gives you everything you need for exam success. We're so confident that they're the best revision books you can buy that if you don't make the grade we will give you your money back!

HERE'S HOW IT WORKS

Register the Letts AS in a Week guide you buy by writing to us within 28 days of purchase with the following information:

- Name
- Address
- Postcode
- Subject of AS in a Week book bought

Please include your till receipt

To make a **claim**, compare your results to the grades below. If any of your grades qualify for a refund, make a claim by writing to us within 28 days of getting your results, enclosing a copy of your original exam slip. If you do not register, you won't be able to make a claim after you receive your results.

CLAIM IF...

You are an AS (Advanced Subsidiary) student and do not get grade E or above.

You are a Scottish Higher level student and do not get a grade C or above.

This offer is not open to Scottish students taking SCE Higher Grade, or Intermediate qualifications.

Registration and claim address:

Letts Success or Your Money Back Offer, Letts Educational, 414 Chiswick High Road, London W4 5TF

TERMS AND CONDITIONS

1. Applies to the Letts AS in a Week series only
2. Registration of purchases must be received by Letts Educational within 28 days of the purchase date
3. Registration must be accompanied by a valid till receipt
4. All money back claims must be received by Letts Educational within 28 days of receiving exam results
5. All claims must be accompanied by a letter stating the claim and a copy of the relevant exam results slip
6. Claims will be invalid if they do not match with the original registered subjects
7. Letts Educational reserves the right to seek confirmation of the level of entry of the claimant
8. Responsibility cannot be accepted for lost, delayed or damaged applications, or applications received outside of the stated registration/claim timescales
9. Proof of posting will not be accepted as proof of delivery
10. Offer only available to AS students studying within the UK
11. SUCCESS OR YOUR MONEY BACK is promoted by Letts Educational, 414 Chiswick High Road, London W4 5TF
12. Registration indicates a complete acceptance of these rules
13. Illegible entries will be disqualified
14. In all matters, the decision of Letts Educational will be final and no correspondence will be entered into

Letts Educational
Chiswick Centre
414 Chiswick High Road
London W4 5TF
Tel: 020 8996 3333
Fax: 020 8743 8390
e-mail: mail@lettsed.co.uk
website: www.letts-education.com

Every effort has been made to trace copyright holders and obtain their permission for the use of copyright material. The authors and publishers will gladly receive information enabling them to rectify any error or omission in subsequent editions.

First published 2000
Reprinted 2001
New edition 2004

British Library Cataloguing in Publication Data
A CIP record for this book is available from the British Library.

ISBN 1 84315 352 1

Cover design by Purple, London

Prepared by *specialist* publishing services, Milton Keynes
Design and project management by Starfish DEPM, London.

Printed in the UK

Letts Educational Limited is a division of Granada Learning Limited, part of Granada plc.

Atomic Structure

How much do you know?

1 Define the terms atomic number, mass number, relative atomic mass and isotope.

2 a) Give the electronic configuration of (i) Al and (ii) Cr.
b) Give the 'electron in box' notation of (i) Ca and (ii) Cu.

3 a) Define the term 'first ionisation energy'.
b) Give the equation for the first ionisation energy of oxygen.

4 Using a mass spectrometer the abundances of the three isotopes of neon, ^{20}Ne, ^{21}Ne and ^{22}Ne, can be determined. They are 90.9%, 0.17% and 8.93% respectively. Calculate the relative atomic mass of neon.

Answers

1 see page 4 **2a)** i) $1s^2\ 2s^2\ 2p^6\ 3s^2\ 3p^1$ ii) (Ar) $3d^5\ 4s^1$ b) i) see diagrams right **3** a) The minimum energy required to remove 1 mole of electrons from each of 1 mole of isolated gaseous atoms. b) $O_{(g)} \rightarrow O^+_{(g)} + e^-$ **4** 20.18

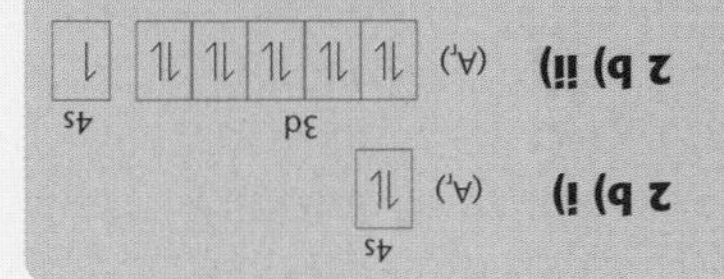

If you got them all right, skip to page 8

Atomic Structure

Learn the key facts

1 The atomic number of an element is the number of protons in the nucleus of an atom. The mass number is the sum of the number of protons and neutrons in an atom. The relative atomic mass of an element is the isotopically-weighted average mass of all the atoms in a sample of the element measured on a scale where ^{12}C = 12.00 g. Isotopes are atoms of the same element but with a different number of neutrons.

2 Electronic configurations can be derived using the diagram below:

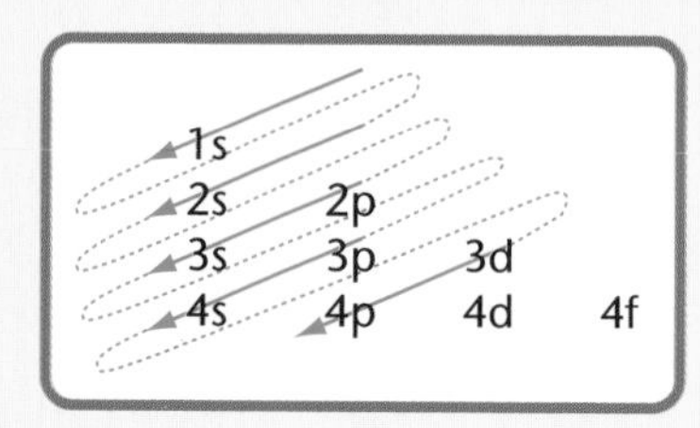

Each orbital can hold a maximum of two electrons. The s, p and d sub-shells each contain 1, 3 and 5 orbitals respectively.

E.g. Al (atomic number 13): is $1s^2\ 2s^2\ 2p^6\ 3s^2\ 3p^1$.

This can also be represented by the 'electron in box' notation:

1s	2s	2p			3s	3p		
↑↓	↑↓	↑↓	↑↓	↑↓	↑↓	↑		

Remember that the electronic configurations of chromium and copper are (Ar) $3d^5\ 4s^1$ and (Ar) $3d^{10}\ 4s^1$ respectively.

3 The first ionisation energy of an element is the minimum energy required to remove one mole of electrons from each of one mole of isolated gaseous atoms.

$$X_{(g)} \rightarrow X^{+}_{(g)} + e^{-}$$

Successive ionisation energies can be defined accordingly i.e. the second ionisation energy is the minimum energy required to remove one mole of electrons from one mole of unipositive cations.

$$X^{+}_{(g)} \rightarrow X^{2+}_{(g)} + e^{-}$$

Successive ionisation energies of an element provide evidence for the characteristic energy levels of orbitals. Consider the graph of the successive ionisation energies of aluminium below:

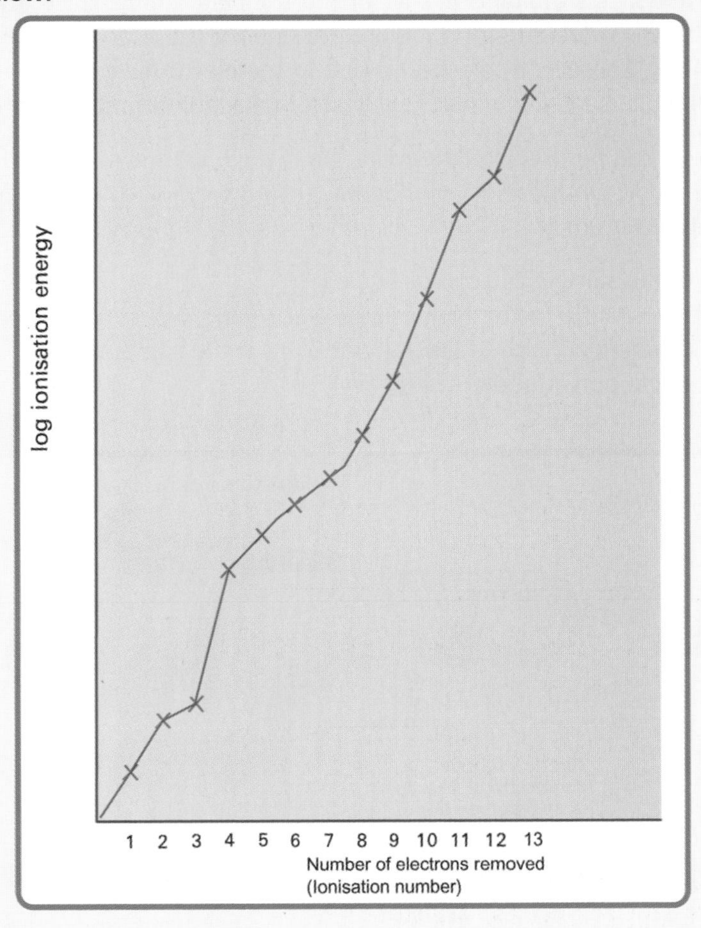

It is important to appreciate that successive ionisation energies of an atom always increase. This is because, as each electron is removed, the size of the ion decreases and so the attraction between the nucleus and the electrons increases.

By referring to the electronic configuration of aluminium (page 4) we can explain the shape of this graph. The first ionisation is the removal of the outer most electron in the p orbital. The next electron to be removed is in the 3s orbital. As a new sub-shell is entered more energy is required, hence the sharp increase in ionisation energy.

The sharp increase in energy between the third and fourth ionisation is due to the 2p orbital being entered. This orbital is in the second shell and hence closer to the nucleus, which is why more energy is required to remove this electron. The increase caused by removing the seventh electron is due to the electronic configuration $1s^2$ $2s^2$ $2p^3$. The half filled p sub-shell is a particularly stable electron arrangement.

4 Mass spectrometry can be used to determine the relative atomic mass of elements, determine the isotopic composition and identify unknown molecules. The five stages in obtaining a mass spectrum are:

- vaporisation of the sample (if solid or liquid);
- ionisation of the sample (using high-energy electrons);
- acceleration (forming a beam of ions all travelling at the same speed);
- deflection (using a powerful electromagnet);
- detection.

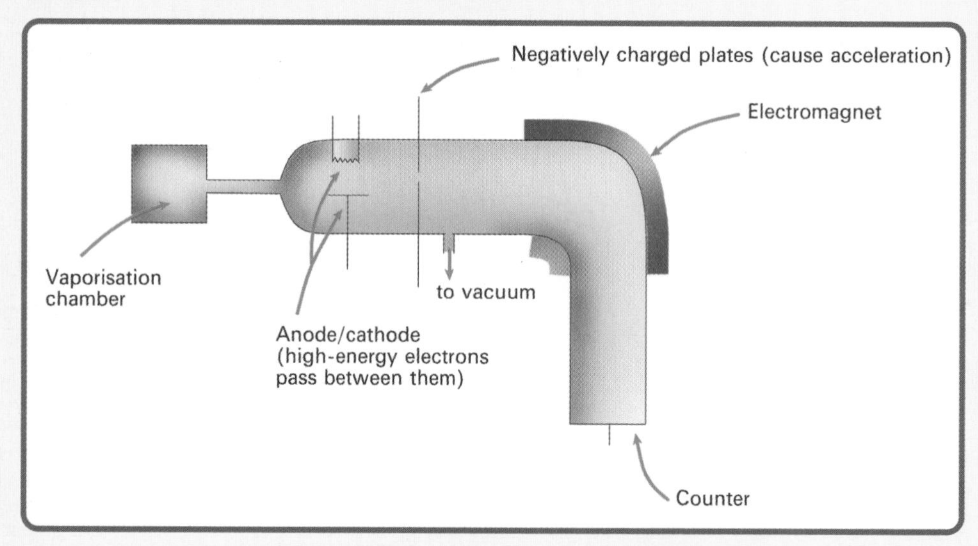

The mass spectrum obtained for lead is shown in the diagram on the next page. Note that the horizontal axis is the ratio of mass to charge. As most ions formed have a charge of +1, this axis can be considered as a mass axis.

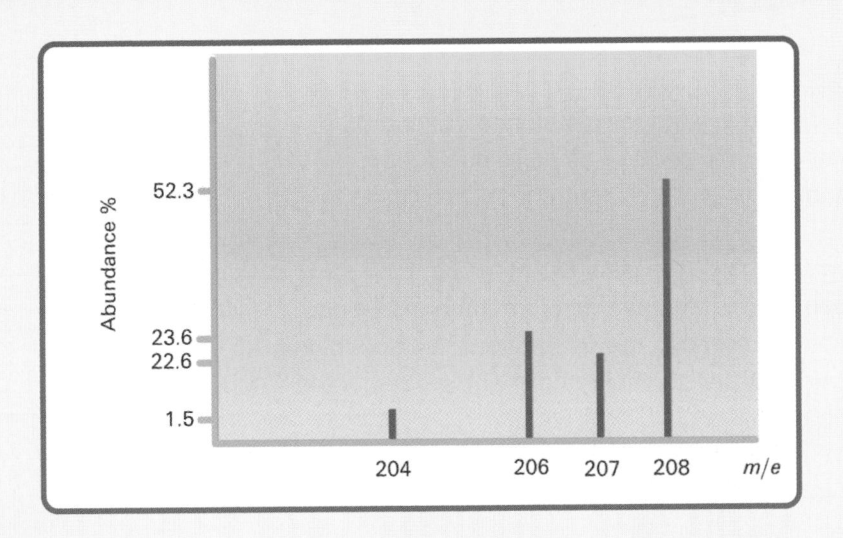

The relative atomic mass is obtained by multiplying the mass of each isotope by the abundance and dividing the total by the number of atoms present. The relative atomic mass of lead is 207.2.

$$\frac{(204 \times 1.5) + (206 \times 23.6) + (207 \times 22.6) + (208 \times 523)}{100}$$

$= 207.2$

Make sure that you:

- *Can define the term relative atomic mass*
- *Understand the existence of isotopes*
- *Can determine the electronic configurations of elements*
- *Can interpret graphs of successive ionisation energy*

Have you improved?

1 This question is about the two isotopes of chlorine ^{35}Cl and ^{37}Cl.

a) Give the electronic configuration of an atom of chlorine.

b) Calculate the number of protons, neutrons and electrons in each isotope.

c) Using your answer to part b) explain the meaning of the term *isotope*.

d) Write an equation representing the first ionisation of chlorine.

e) Write an equation representing the second ionisation of chlorine.

f) Why is the second ionisation energy of chlorine greater than the first?

Learn the sequence of orbital filling

Consider the mass number and the atomic number

Learn the definition of isotope!

Think about the number of protons and electrons

2 The mass spectrum of the element boron is shown below:

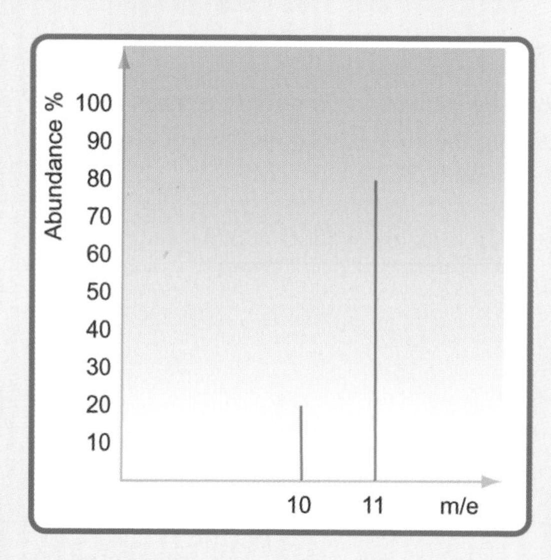

a) What is the percentage abundance of each isotope?

b) What does m/e represent?

c) What will be the charge on the boron ions produced in the spectrometer?

d) Calculate the relative atomic mass of boron.

Read off from the graph

Consider the ionisation stage

See example on page 7

Formulae and Equations

How much do you know?

DAY 1

1 a) Calculate the empirical formula of the following compounds:
   i) A substance containing 0.31 g of phosphorus and 1.065 g of chlorine.
   ii) A hydrocarbon containing 85.7% by mass of carbon.
b) Calculate the molecular formula of the hydrocarbon in ii) above, given that its molecular mass is 112.

2 Balance the following equations:
a) $LiOH + H_2SO_4 \rightarrow Li_2SO_4 + H_2O$
b) $CH_4 + O_2 \rightarrow CO_2 + H_2O$
c) $Fe + H_2O \rightarrow Fe_3O_4 + H_2$

3 Write ionic equations for the following reactions:
a) $NaOH_{(aq)} + HCl_{(aq)} \rightarrow NaCl_{(aq)} + H_2O_{(l)}$
b) $FeCl_{3(aq)} + 3NaOH_{(aq)} \rightarrow Fe(OH)_{3(s)} + 3NaCl_{(aq)}$
c) $ZnO_{(s)} + H_2SO_{4(aq)} \rightarrow ZnSO_{4(aq)} + H_2O_{(l)}$

4 Write the ideal gas equation in symbolic form.

Answers

1a) i) PCl_3 ii) CH_2 b) C_8H_{16}
2a) $2LiOH + H_2SO_4 \rightarrow Li_2SO_4 + 2H_2O$ b) $CH_4 + 2O_2 \rightarrow CO_2 + 2H_2O$
c) $3Fe + 4H_2O \rightarrow Fe_3O_4 + 4H_2$
3a) $H^+_{(aq)} + OH^-_{(aq)} \rightarrow H_2O_{(l)}$ b) $Fe^{3+}_{(aq)} + 3OH^-_{(aq)} \rightarrow Fe(OH)_{3(s)}$
c) $2H^+_{(aq)} + O^{2-}_{(aq)} \rightarrow H_2O_{(l)}$
4 $PV = nRT$ or $PV = m/M\ RT$

If you got them all right, skip to page 12

Formulae and Equations

Learn the key facts

1 The empirical formula of a compound is the simplest whole-number ratio of the atoms in a molecule. The molecular formula is the actual number of atoms present and is a multiple of the empirical formula.

Suppose we want to find the empirical formula of a hydrocarbon containing 80% carbon and 20% hydrogen. (Percentage composition is treated in the same way as mass composition.)

We use these steps to determine the empirical formula:

	Element	**C**	**H**
Step 1	Mass	80	20
Step 2	÷ RAM(A_r)	80/12 t 6⅔	20/1 t 20
Step 3	÷ Smallest	1	3

Thus the empirical formula is CH_3.

To determine the molecular formula the molecular mass needs to be known.

Suppose we want to find the molecular formula of the above hydrocarbon, given that its molecular mass is 30.

- Each empirical formula unit has a mass of 15 ($15 = 12 + (3 \times 1)$).
- The number of empirical formula units in the molecular formula is found by dividing the molecular mass by the mass of each empirical formula unit.
- There are two units in the molecular formula ($30 \div 15 = 2$).
- Hence the molecular formula becomes C_2H_6 (the number of each atom present is multiplied by 2).

2 Chemical equations must be balanced. This means that the number of each type of atom must be the same on both sides of the equation.

If they aren't, then you have to put numbers in front of the formulae containing that atom to make them the same, as shown on page 11.

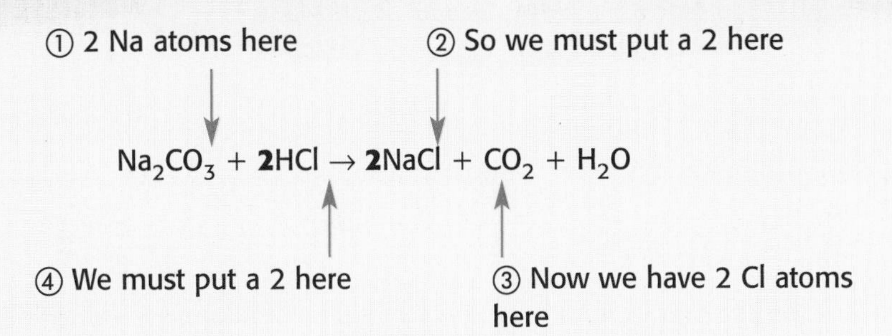

3 Ionic equations can be written if the following rules are followed:

- Write out the balanced symbol equation.
- Separate out each substance into its component ions (except for any liquids and solids that are products).
- Cancel any common ions.

For example, the ionic equation for the reaction of silver nitrate solution with dilute hydrochloric acid, according to the steps above, is:

1 $AgNO_{3(aq)} + HCl_{(aq)} \rightarrow AgCl_{(s)} + HNO_{3(aq)}$

2 $Ag^+{}_{(aq)} + NO_3^-{}_{(aq)} + H^+{}_{(aq)} + Cl^-{}_{(aq)} \rightarrow AgCl_{(s)} + H^+{}_{(aq)} + NO_3^-{}_{(aq)}$

3 $Ag^+{}_{(aq)} + \cancel{NO_3^-{}_{(aq)}} + \cancel{H^+{}_{(aq)}} + Cl^-{}_{(aq)} \rightarrow AgCl_{(s)} + \cancel{H^+{}_{(aq)}} + \cancel{NO_3^-{}_{(aq)}}$

The ionic equation is therefore: $Ag^+{}_{(aq)} + Cl^-{}_{(aq)} \rightarrow AgCl_{(s)}$

4 The ideal gas equation is

$PV = nRT$

where P is the pressure (in Pa)
V is the volume (in dm^3)
n is the number of moles being considered
R is the universal gas constant (8.31J K^{-1} mol^{-1})
T is the temperature (in Kelvin)

NB: $n = m/M$ where m is the mass of the sample of gas (in grams) and M is the relative molecular mass of the gas.

Make sure that you:

- *Can determine the empirical formula*
- *Can use empirical formulae to determine molecular formulae*
- *Can balance symbol equations*
- *Are able to write ionic equations*
- *Can use the ideal gas equation*

Formulae and Equations

Have you improved?

1 The percentage composition by mass of a carboxylic acid was found to be:

C: 40.7% H: 5.1% O: 54.2%

a) Calculate the empirical formula of the carboxylic acid.
b) Determine the molecular formula of the carboxylic acid given that the relative molecular mass is 118.

Learn the three steps (page 10)

See example on page 10

2 Sodium carbonate solution reacts with hydrochloric acid to form sodium chloride, water and carbon dioxide.
a) Write a balanced symbol equation for this reaction.
b) Write an ionic equation for this reaction.

Check you know the correct formula of each species

See section 3 on page 11

3 A 0.0005 m^3 sample of a gas weighed 1.3g at 25°C and atmospheric pressure (101 325 Pa). Using the ideal gas equation calculate the relative molecular mass of the gas. ($R = 8.314$ J K^{-1} mol^{-1}).

Remember $n = m/M$

Moles

15 mins
Time Yourself

How much do you know?

1 a) What is the numerical value of Avogadro's number?
b) What do you understand this number to mean?

2 a) Calculate the mass of iron (III) chloride produced when 5.6 g of iron reacts with excess chlorine according to the equation:

$$2Fe_{(s)} + 3Cl_{2(g)} \rightarrow 2FeCl_{3(s)}$$

b) Determine the mass of sodium that would be required to form 0.31 g of sodium oxide by the following reaction:

$$4Na_{(s)} + O_{2(g)} \rightarrow 2Na_2O_{(s)}$$

3 Look at this equation:

$$Na_2CO_{3(s)} + 2HCl_{(aq)} \rightarrow 2NaCl_{(aq)} + CO_{2(g)} + H_2O_{(l)}$$

a) Calculate the concentration of a sodium carbonate solution, given that 25 cm^3 reacted with 22.6 cm^3 of 0.2 M HCl.
b) What mass of sodium carbonate would need to be dissolved in 250 cm^3 of distilled water to produce a solution of this concentration?

4 Calculate the volume of ammonia produced if $\frac{1}{4}$ mole of nitrogen reacts with $\frac{3}{4}$ mole of hydrogen according to the equation below, 1 mole of gas occupies 24 dm^3 at 298 K.

$$N_2 + 3H_2 \rightarrow 2NH_3$$

Answers

1a) 6.02×10^{23} b) The number of particles present in a mole.
2a) 16.25 g b) 0.23 g
3a) 0.0904 mol dm^{-3} b) 2.40 g
4 12 dm^3

If you got them all right, skip to page 16

Spend no more than 30 mins on this topic

DAY 1 2 3 4 5 6 7

Moles

Learn the key facts

1 Avogadro's constant or number (N_A), is 6.02×10^{23}. One mole of any substance has 6.02×10^{23} particles in it. The mass of one mole of a substance is the same as its relative atomic (or molecular) mass. The mole is a useful concept when doing calculations involving equations because equations tell you how many moles of each substance react/are formed.

2 To work out reacting masses from chemical equations consider this equation:

$$CaCO_{3(s)} + 2HCl_{(aq)} \rightarrow CaCl_{2(aq)} + H_2O_{(l)} + CO_{2(g)}$$

To find the mass of calcium chloride formed if 50 g of $CaCO_3$ reacts, it is necessary to work in moles (the equation tells us the ratio of reacting moles).

These general rules should help you to solve mole questions.

Step 1 Calculate the number of moles of the substance that you have most information about using one of these three formulae:

$$\text{Number of moles (of solid)} = \frac{\text{mass}}{\text{RAM (Ar) or RMM (Mr)}}$$

$$\text{Number of moles (in solution)} = \frac{\text{volume (cm}^3) \times \text{concentration (mol dm}^{-3})}{1000}$$

$$\text{Number of moles (of gas)} = \frac{\text{volume (cm}^3)}{24000} \text{ (at r.t.p.)}$$

Step 2 Look at the ratio of the reactant and products (from the equation) to determine how many moles of the substance you are interested in are present.

Step 3 Usually it is necessary to calculate the mass, volume or concentration of the substance you are interested in. To do this, rearrange one of the above formulae.

For the above equation we would proceed as follows:

Step 1 The number of moles of calcium carbonate is 50/100 = 0.5.
Step 2 The ratio of moles of calcium carbonate to calcium chloride is 1:1.
Step 3 The mass of calcium chloride is found by multiplying the number of moles present (0.5) by the RMM of calcium chloride (111) = 55.5 g.

3 Titration calculations are used to calculate concentrations.

Calculate the concentration of sodium hydroxide solution given that 25 cm^3 of the solution reacts with 30 cm^3 of 0.1 M sulphuric acid.

The equation for the reaction is:

$$2NaOH_{(aq)} + H_2SO_{4(aq)} \rightarrow Na_2SO_{4(aq)} + 2H_2O_{(l)}$$

Step 1 Number of moles of H_2SO_4 is $(30 \times 0.1)/1000 = 0.003$.
Step 2 Ratio is 1:2 hence the number of moles of NaOH reacting is 0.006.
Step 3 So the concentration of NaOH is $(0.006 \times 1000)/25 = 0.24$ M.

4 Avogadro's Law states that equal volumes of gases contain an equal number of moles (provided that temperature and pressure are constant). We can use Avogadro's law to determine gas volumes in equations.

E.g. Calculate the volume of the products when 20 cm^3 of ethane (C_2H_6) is burnt in air:

The equation for the reaction is:

$$2C_2H_{6(g)} + 7O_{2(g)} \rightarrow 4CO_{2(g)} + 6H_2O_{(g)}$$

According to the equation, 2 moles of ethane form 4 moles of carbon dioxide and 6 moles of water vapour. Therefore 20 cm^3 of ethane forms 40 cm^3 of carbon dioxide and 60 cm^3 of water vapour. Therefore the total volume of the products is 100 cm^3.

Make sure that you:

- *Know Avogadro's number*
- *Can determine reacting masses*
- *Understand titration calculations*
- *Can use Avogadro's Law*

Moles

Have you improved?

1 a) Calculate the number of moles of atoms in each of the following:
i) 1 mole of Cl_2
ii) ½ mole of S_8
iii) 2 moles of CH_4

b) Using Avogadro's number determine the total number of atoms present in each of the following:
i) 1 mole of carbon atoms
ii) 1 mole of oxygen molecules
iii) ½ mole of water molecules

Look at the formula

2 A 3.0 g sample of marble neutralised 24 cm^3 of 2 M HCl. Calculate the percentage of calcium carbonate in the marble.

Follow the mole calculation rules

3 18.20 g of a metal hydroxide XOH (where X is a Group 1 metal) was dissolved in 250 cm^3 of distilled water. 25 cm^3 aliquots of this solution were titrated with 2.10 M HCl. The average titre was 15.45 cm^3. Determine the identity of X.

Calculate the RMM of XOH

4 Calculate the volume of hydrogen produced when 0.20 moles of zinc are placed in excess hydrochloric acid. NB: One mole of gas at 298 K and 1 atm pressure occupies 24 dm^3.

$$Zn_{(s)} + 2HCl_{(aq)} \rightarrow ZnCl_{2(aq)} + H_{2(g)}$$

Look at the ratio of moles of zinc to hydrogen

Structure and Bonding

How much do you know?

1 Explain the term ionic bond.

2 Which ionic bond would you expect to be stronger: NaF or NaI?

3 Which cation has the highest charge to size ratio: Be^{2+}, Ba^{2+} or Li^{+}?

4 Why do ionic substances conduct electricity when molten?

5 Describe metallic bonding.

6 Why are metals malleable?

7 Explain the term covalent bond.

8 Which has the strongest covalent bond: Cl_2 or I_2?

9 Which of the following molecules are polar: CH_3Cl, CO_2, H_2?

10 Which of these molecules show H-bonding: HCl, NH_3, CH_3OH, HCN?

11 What kind of inter-molecular forces are present in oxygen?

12 Why does HF have a higher than expected boiling point?

13 Explain why the molecule CH_4 is tetrahedral.

14 Explain why the bond angle in NH_3 is less than that in CH_4.

15 What type of structure is present in SiO_2?

16 Name the process by which a solid turns directly into a gas on heating.

Answers

1 attraction between opposite ions formed by electron transfer **2** Na–F **3** Be^{2+} **4** They contain mobile ions. **5** attraction between metal cations and delocalised electrons **6** When metal lattice deforms, electrons flow into new 'spaces' to maintain bonding. **7** attraction of two nuclei for shared pair of electrons **8** Cl_2 **9** CH_3Cl **10** CH_3OH, NH_3 **11** van der Waal's. **12** Contains strong hydrogen bonding. **13** 4 electron pairs repel each other equally. **14** Lone pair repels more than bond pairs. **15** giant covalent **16** sublimation

If you got them all right, skip to page 23

Structure and Bonding

Learn the key facts

1 An ionic bond is the electrostatic attraction between two oppositely charged ions formed as a result of electron transfer. Metals readily lose electrons (forming cations) and non-metals readily accept electrons (forming anions) to achieve a full octet of electrons, e.g. Na^+, Ca^{2+}, Cl^-, O^{2-}.

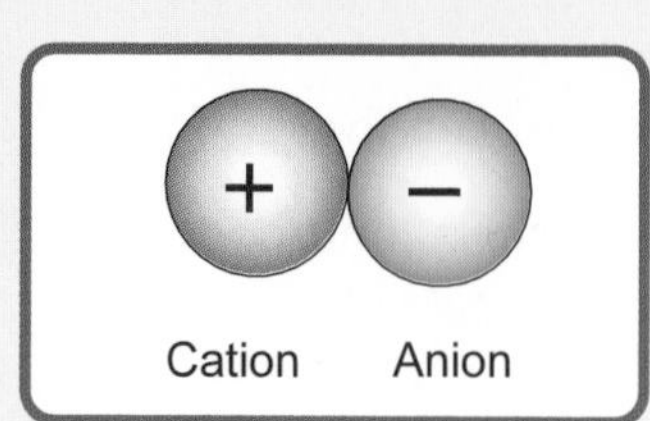

2 Ionic bond strength increases with ionic charge and decreases with ionic size (separation). Their ratio is called surface charge density, and thus ionic bond strength increases with charge density. E.g. bonding in NaF is stronger than NaBr because F^- has a larger charge density than Br^-; bonding in MgF_2 is stronger than NaF as Mg^{2+} has greater charge density than Na^+.

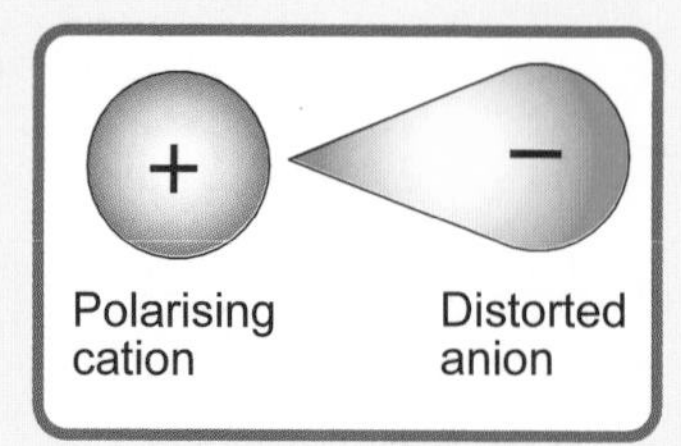

3 Cations with very high charge densities (e.g. Al^{3+}, Be^{2+} and Li^+) polarise electron density from anions towards themselves. The anion is distorted from a spherical shape so that some electron density lies between the ions and the bond has some covalent character, e.g. Al_2O_3, $MgCl_2$, LiBr. Large anions are readily polarisable due to weaker nuclear attraction for the outer electrons, (e.g. I^-, Br^-, Se^{2-}).

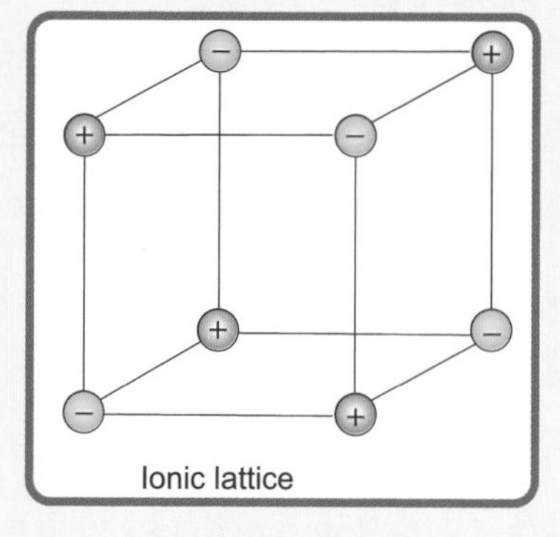

4 Ionic solids consist of a regular lattice of alternating cations and anions. The lattice is held together by many strong ionic bonds requiring a lot of energy to break down, leading to high melting points. When molten or in solution, ions are free to move and conduct an electrical current. Ionic solids dissolve in H_2O to form an aqueous solution *if* the energy required to break down the lattice is compensated for by the energy released during hydration, where ions are surrounded by polar water molecules, attracted to the charged ions.

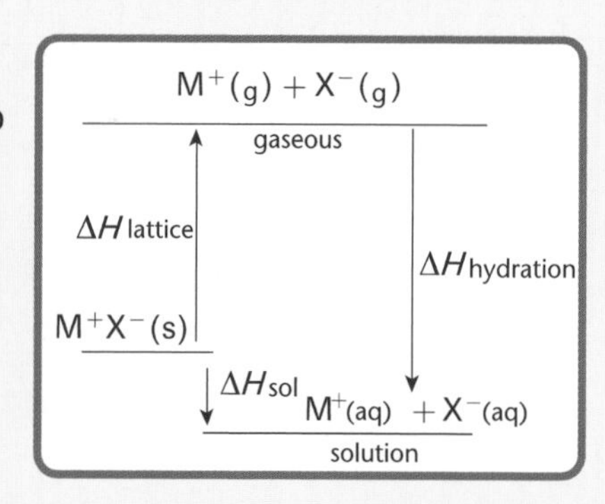

5 A metallic bond is the electrostatic force of attraction between metal cations (arranged in a regular lattice) and the delocalised, valence electrons surrounding them. Bond strength depends on the number of delocalised electrons and cation surface charge density, e.g. the bonding in Mg (+2 cations and 2 valence e^-'s) is stronger than in Na (+1 cations and 1 valence e^-).

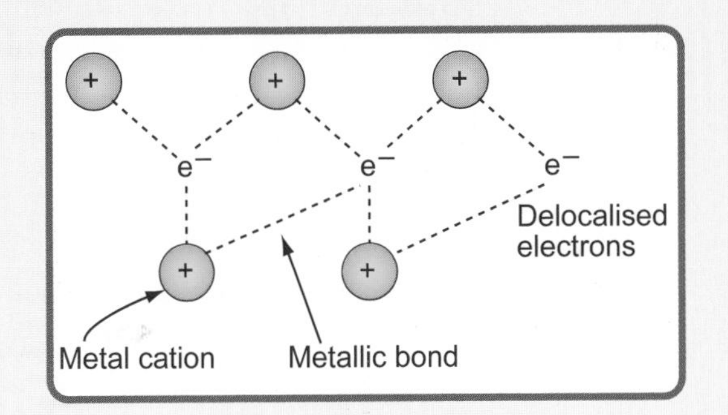

6 As all the metallic bonds must be broken, metals have relatively high melting and boiling points. Metals are good conductors of electricity because the delocalised electrons are free to move in all directions. They are malleable and ductile because as the cation lattice distorts, the delocalised electrons 'flow' into the new 'spaces' to maintain bonding.

7 A covalent bond is the electrostatic force of attraction between two nuclei and a *shared* pair of electrons between them. Covalent bonding occurs between non-metal atoms which cannot easily gain or lose electrons and which achieve a full octet by sharing electrons (illustrated by dot and cross diagrams). E.g. in H_2O, CH_4, diamond and HCl.

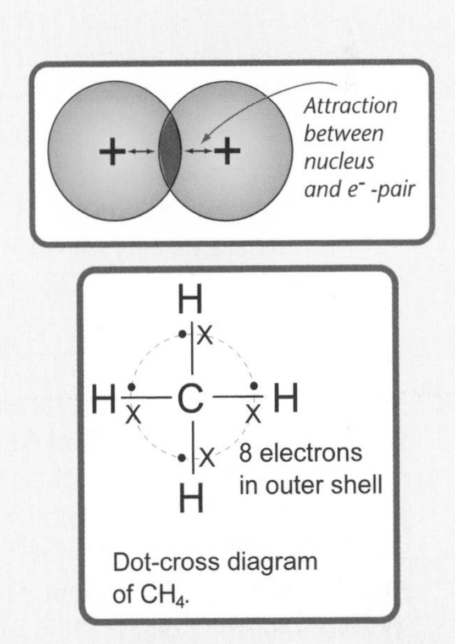

Dot-cross diagram of CH_4.

8 Small atoms form strong covalent bonds because their nuclei are close to the bonding electrons resulting in a stronger force of attraction, e.g. the covalent bond in Cl_2 is stronger than that in I_2. Dative covalent bonds occur when one atom provides both electrons, e.g. the ammonium ion $H_3N{:}\rightarrow H^+$.

9 Electron density in covalent bonds is distorted towards the more electronegative atom causing a polar-covalent bond (δ^+, δ^-). Molecules with polar bonds often have an overall dipole. If they are symmetrical, bond dipoles will cancel out and show no dipole.

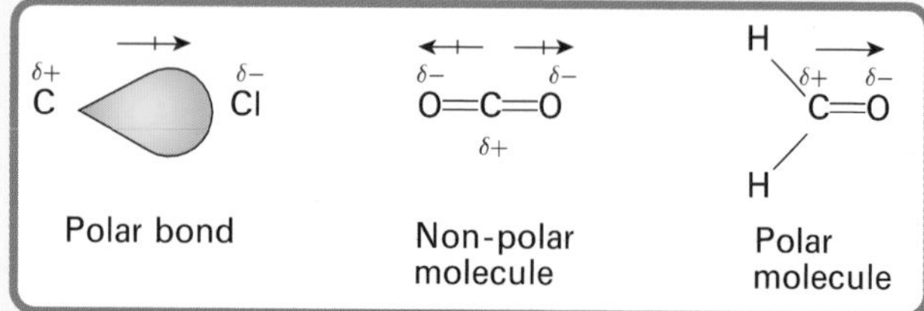

10 Covalent substances can have simple-molecular structures, consisting of separate molecules held together by weak intermolecular forces.

- van der Waals forces are the electrostactic force of attraction between temporary and induced dipoles. They are relatively weak.
- Permanent dipole–dipole attractions exist between the δ^+ and δ^- regions of polar molecules, and are stronger than vdW forces, e.g. carbonyls $CHCl_3$.
- Hydrogen bonds exist between the lone pair of a N, O, or F atom (bonded to H) and the δ^+ H atom bonded to N, O or F on a neighbouring molecule.
  H-bonds are usually the strongest intermolecular forces, e.g. H_2O, NH_3, HF.

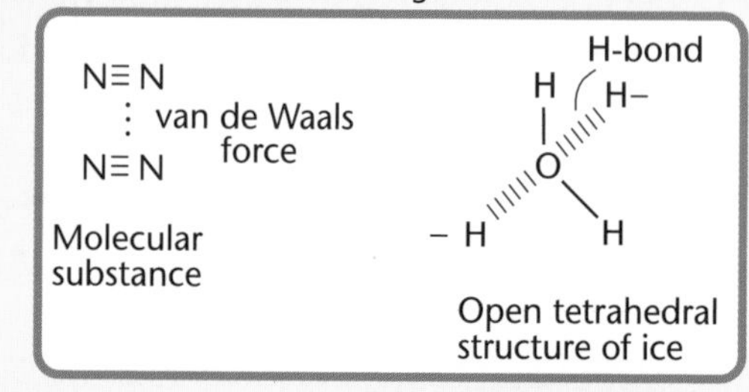

11 Most molecular substances have relatively low melting and boiling points as only weak van der Waals forces are broken, (CO_2, N_2). As molecules become larger and therefore more polarisable, van der Waals forces become stronger and so melting and boiling points increase, e.g. I_2 has a highter melting point than F_2. Because usually there are no mobile, charged particles, molecular substances do not conduct electricity. Unless they have a significant, permanent dipole they are not usually soluble in water as they cannot interact with the polar H_2O molecules, e.g. alkanes.

12 Relatively strong hydrogen bonds lead to a significant increase in melting and boiling points, e.g. the boiling point of H_2O is greater than that of H_2S *despite* being smaller. Hydrogen bonding also leads to improved solubility in water as molecules can hydrogen bond with H_2O molecules, e.g. alcohols are soluble in water, unlike alkanes. Extensive hydrogen bonding is also responsible for the open, low-density structure of ice.

13 Molecular shapes are based on the number of electron pairs on the central atom, which repel each other to form geometric shapes. Shapes can be predicted as follows, e.g. for the NH_3 molecule:

Step		Example
1	1 for each valence electron of the central atom.	N = +5
2	1 for each atom singly bonded.	3 H = +3
3	+1 per negative charge and −1 per positive charge.	no charge = 0
		total = 8
4	Halve the total to give number of pairs and base shape.	(4 = tetrahedral)
5	Assign bond pairs and lone pairs to give observed shape.	(3 bond pairs, 1 lone pair = pyramidal)

Other shapes are given below:

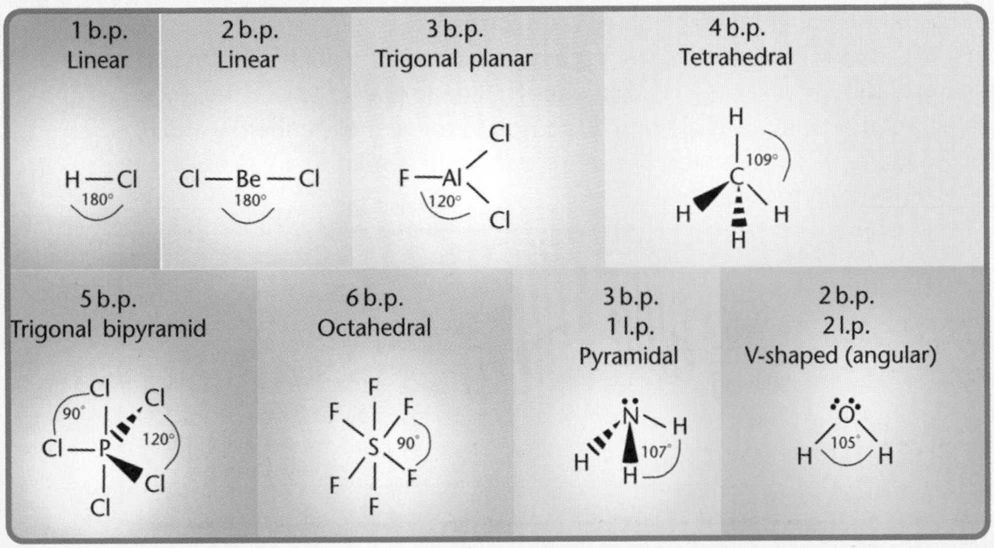

1 b.p. Linear	2 b.p. Linear	3 b.p. Trigonal planar	4 b.p. Tetrahedral
H—Cl 180°	Cl—Be—Cl 180°	F—Al(Cl)(Cl) 120°	CH_4 109°

5 b.p. Trigonal bipyramid	6 b.p. Octahedral	3 b.p. 1 l.p. Pyramidal	2 b.p. 2 l.p. V-shaped (angular)
PCl_5 90° 120°	SF_6 90°	NH_3 107°	H_2O 105°

DAY 2

14 Lone pairs (l.p.) reduce bond angles as they repel more strongly than bond pairs (b.p.). E.g. the shapes of CH_4 (tetrahedral), NH_3 (pyramidal) and H_2O (bent) are all based on 4 electron pairs but the bond angle decreases from approximately 109° to 107° to 105°.

15 Atoms can covalently bond to from a lattice referred to as a giant covalent structure, e.g. SiO_2 and diamond. As all the strong covalent bonds must be broken, requiring a large amount of energy, these substances have high melting points. They are not soluble in water because they contain no ions or hydrogen bonds. They do not usually conduct electricity because they contain no mobile, charged particles. The exception is graphite, which consists of layers of covalently bonded carbon atoms with delocalised electrons between them. An electrical current can therefore flow *between* layers.

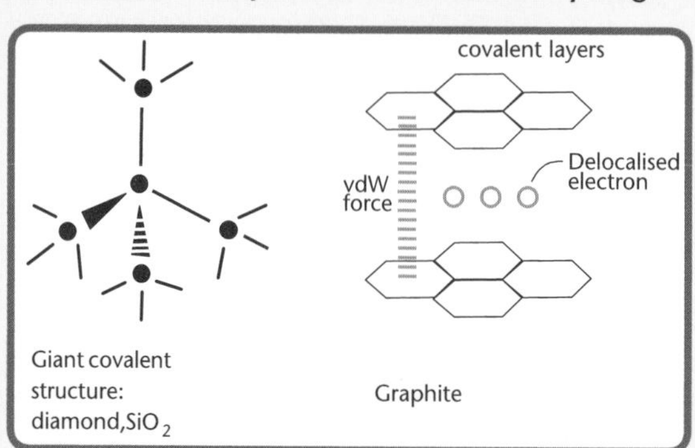

16 Solids contain particles arranged in a regular lattice which vibrate about a fixed point. As they are heated, particles gain kinetic energy and vibrate more quickly until they have enough energy to break the forces binding them and melting occurs. Liquids still have a localised, regular structure but particles can flow past one another. Eventually, particles gain enough kinetic energy to boil, i.e. they break all attractions to each other and are free to move independently and randomly as a gas. Some solids change directly from a solid to a gas on heating – this is called sublimation.

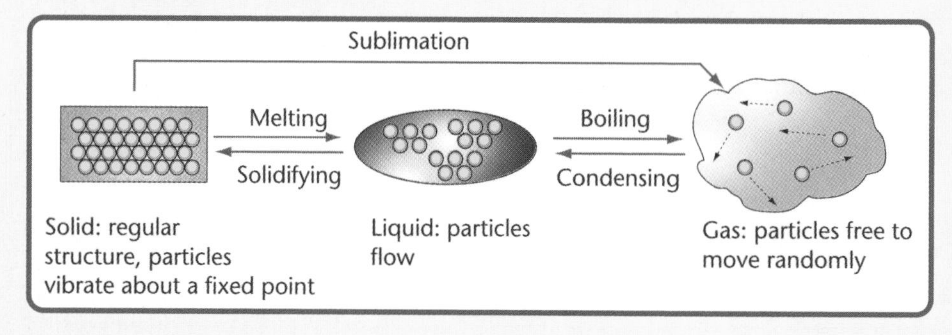

Make sure you are familiar with the following key terms:

- *ionic bond*
- *surface charge density*
- *polarisation*
- *ionic lattice*
- *hydration*
- *metallic bonding*
- *delocalised electrons*
- *covalent bond*
- *dative bond*
- *dipole*
- *van der Waals forces*
- *dipole–dipole attractions*
- *hydrogen bonding*
- *physical properties of ionic and covalent materials*
- *molecular shapes*
- *structure of solids, liquids and gases*
- *melting, boiling and sublimation*

Have you improved?

1 Phosphorous trichloride, PCl_3, is a simple molecular substance.

a) Predict and explain the shape of the covalent molecule PCl_3.

b) Suggest and explain the Cl–P–Cl bond angle.

c) Why does PCl_3 has a relatively low melting point? Explain what happens during melting in terms of molecular motion.

How many electron pairs on P?

Lone pair present

What type of bonds are broken?

2 The elements of group 6 all form hydrides of the general formula H_2X.

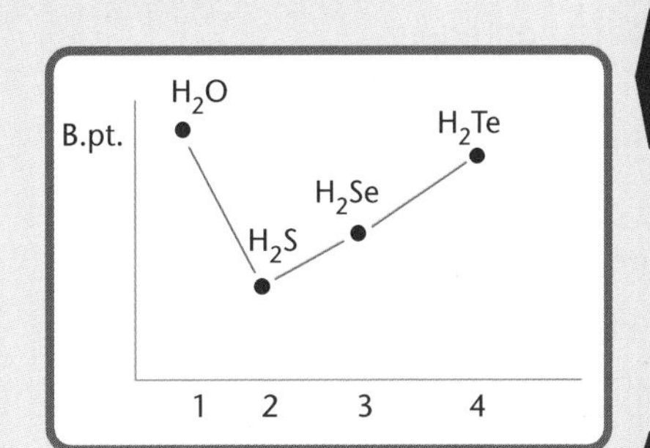

a) Draw a dot cross diagram for H_2S.

b) Explain the variation in boiling point of the group 6 hydrides shown in the graph.

c) Explain why ice is less dense than water.

Show all electrons around O atom

Unique bonding in H_2O

Ice molecules form 4 bonds

3 Chlorine and Iodine are both in group 7.

a) Explain why chlorine contains a stronger covalent bond than iodine.

b) Both elements form ionic salts with potassium. Explain why these salts dissolve in water and their solutions conduct electricity.

c) Lithium iodide is said to have covalent character. Explain what is meant by covalent character and how it occurs.

Atomic size
Free ions interact with H_2O

Li^+ small I^- large

The Periodic Table

How much do you know?

1 The elements in the periodic table are arranged in order of increasing

____________ ____________.

2 a) State the trend across Period 3 of each of the following:
   i) atomic radius
   ii) ionisation energy
   iii) electron affinity.
b) Explain each of these trends.

3 Which one of the following has the highest melting point?
a) sodium, magnesium or aluminium
b) sulphur, phosphorus or chlorine

4 Magnesium has a metallic structure. Explain why magnesium conducts electricity.

Answers

1 atomic number
2a) i) Decreases across the period. ii) Increases across the period (becomes more endothermic).
iii) Becomes more exothermic across the period.b) See page 25.
3a) aluminium b) sulphur
4 Metallic structures contain delocalised (free moving) electrons.

If you got them all right, skip to page 27

Learn the key facts

1 The periodic table contains the elements arranged in order of increasing atomic number (proton number).

2 Electronic configurations (see page 4) are a periodic function. We therefore expect that physical properties connected with electronic arrangement will themselves exhibit periodicity.

Across a period the atomic radius decreases because the addition of an extra proton pulls in the outer electrons more strongly. The extra electron added causes little increase in shielding because it is added to the same shell. This decrease in radius causes the ionisation energy to increase and the electron affinity to become more exothermic.

Down a group, the atomic radius increases because the extra shell increases the shielding of the outer electrons from the nucleus. The ionisation energy, therefore, decreases and the electron affinity becomes more endothermic.

The periodicity of some properties such as melting point is not so straightforward. This is because many properties depend upon the structure and bonding rather than the properties of isolated gaseous atoms.

3 The graph represents the melting point of the Period 3 elements.

The periodicity in melting points can be explained by considering the structure and bonding of the elements. Sodium, magnesium and aluminium all have metallic lattice structures.

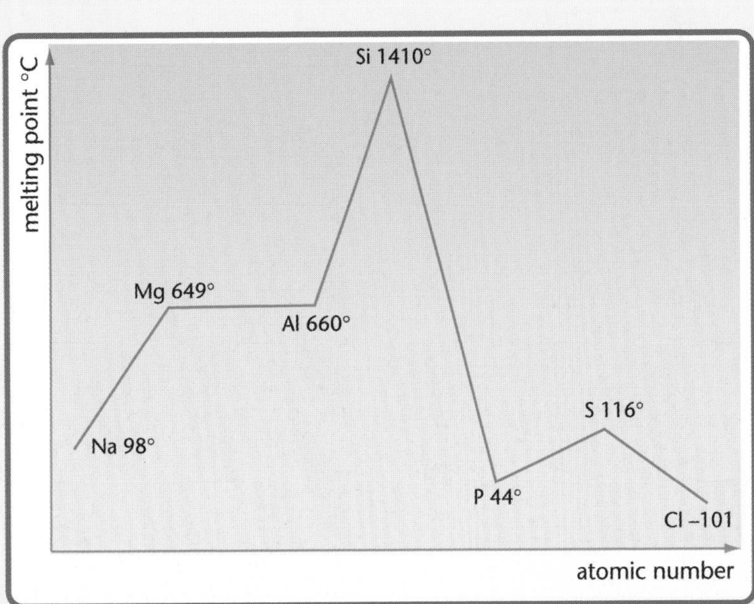

The strength of metallic bonding within them increases across the period because the charge density of the metal cations increases. This means that the melting point increases too. Silicon has the highest melting point of the period due to its giant covalent (diamond) structure. Phosphorus, sulphur and chlorine all have relatively low melting points due to their simple covalent structure. The bigger the molecule (Cl_2, P_4, S_8) the greater the van der Waals bonding and hence the higher the melting point.

4 The trend in electrical conductivity can also be explained by considering the structure and bonding of the elements. Sodium, magnesium and aluminium are all good conductors of electricity becaue they have a metallic lattice. Silicon is a semi-conductor and phosphorus, sulphur and chlorine do not conduct electricity due to their simple covalent structures.

Make sure that you:

Know the periodic trends in atomic radius, ionisation energy and electron affinity

Can explain the periodic trends in melting point

Can explain the periodic trends in electrical conductivity

Have you improved?

1 The data in the table below is about three elements in period 3 that are adjacent to each other.

Element	Atomic radius (nm)	First ionisation energy ($kJmol^{-1}$)	Electrical conductivity
X	0.130	578	good
Y	0.118	789	very weak
Z	0.110	1012	non-conductor

a) How does the atomic radius vary as you go across the period?

b) Use your answer to part a) to explain the trend in first ionisation energy.

c) Using your knowledge of the periodic table, suggest identities for X, Y and Z.

Look at the data

Learn the definition (chapter 1)

Consider electrical conductivity

2 Sodium and chlorine are both elements in period 3 but they have very different properties. Sodium is a solid element that conducts electricity. Chlorine is a gas that is a non-conductor.

a) With reference to structure, explain why sodium is a conductor of electricity.

b) With reference to structure, explain why chlorine is a gas at room temperature.

c) Explain why chlorine gas does not conduct electricity.

Sodium has a metallic structure

Consider intermolecular forces

Chlorine is a covalent molecule

Introduction to Oxidation and Reduction

How much do you know?

1 a) Define the term oxidation number.
b) Calculate the oxidation number of the element in red in each of the species below:
i) Cl_2 ii) NaCl iii) H_2O_2 iv) MnO_4^-

2 Define the terms below, in terms of electrons.
a) oxidation
b) reduction
c) oxidising agent
d) reducing agent

3 Complete the following half-equations:
a) $Cl^- \rightarrow Cl_2$
b) $Fe^{3+} \rightarrow Fe^{2+}$
c) $MnO_4^- \rightarrow Mn^{2+}$

4 Combine the following two half-equations to give the overall equation.

$$Cr_2O_7^{2-} + 14H^+ + 6e^- \rightarrow 2Cr^{3+} + 7H_2O$$

$$2I^- \rightarrow I_2 + 2e^-$$

Answers

1a) See section 1, page 29. b) i) 0 ii) −1 iii) −1 iv) +7
2a) loss of electrons b) gain of electrons c) a substance that is itself reduced
d) a substance that is itself oxidised
3a) $2Cl^- \rightarrow Cl_2 + 2e^-$ b) $Fe^{3+} + e^- \rightarrow Fe^{2+}$ c) $MnO_4^- + 8H^+ + 5e^- \rightarrow Mn^{2+} + 4H_2O$
4) $Cr_2O_7^{2-} + 6I^- + 14H^+ \rightarrow 2Cr^{3+} + 3I_2 + 7H_2O$

If you got them all right, skip to page 32

Introduction to Oxidation and Reduction

Spend no more than **30 mins** on this topic

Learn the key facts

1 The oxidation number is the number of electrons that need to be added to a positive ion or taken away from a negative ion to give a neutral atom. In covalent compounds the electrons are assumed to go to the most electronegative atom. The following rules allow us to determine the oxidation number of elements in inorganic species:

- The oxidation number of all elements is 0.
- The sum of the oxidation numbers of the elements in a species is equal to the charge on an ion (or zero for a neutral molecule).
- The oxidation number of group 1 and group 2 elements is always +1 and +2 respectively.
- In covalent molecules where one atom is more electronegative than the other, the more electronegative element will have a negative oxidation number.
- The oxidation number of oxygen is always –2 except in peroxides $(O_2)^{2-}$ and in oxygen difluoride (F_2O) where the oxidation state of oxygen is –1 and +2 respectively (potassium, rubidium and caesium each form a 'superoxide' MO_2 in which the oxidation number of oxygen is $-\frac{1}{2}$).

In the examples below, the oxidation state of the highlighted element in each species is determined.

$Cr_2O_7^{2-}$ Each oxygen has an oxidation number of –2 (total –14). The overall charge on the ion is –2 so the two chromiums must therefore have a combined oxidation number of (+12) because $-14 + 12 = -2$. The oxidation number of chromium is therefore +6.

$NaVO_3$ Sodium (a group 1 metal) has an oxidation state of +1 and the three oxygens a total of –6. The molecule is neutral and so the vanadium must have an oxidation number of +5 because $+1 +5 -6 = 0$.

2 Oxidation can be defined (in terms of electron transfer) as the loss of electrons from a species and reduction can be defined as the gain of electrons by a species. The acronym OILRIG ('Oxidation Is Loss of electrons, Reduction Is Gain of electrons') can be helpful in remembering this. It follows from these definitions that an oxidising agent (electron acceptor) must itself be reduced (because in order for an oxidation to occur a reduction must also occur) and similarly a reducing agent (electron donor) must itself be oxidised.

3 The action of both oxidising and reducing agents can be represented in the form of a half-equation. The following rules allow us to construct balanced half-equations.

- Write down the formula of the reactant and product (you will be told these).
- Add sufficient hydrogen ions to react with any oxygen atoms present, forming the appropriate number of water molecules.
- Add sufficient electrons to one side of the equation so that the charges balance.

E.g. The half-equation for the reduction of the dichromate ion ($Cr_2O_7^{2-}$) to the chromium (III) ion (Cr^{3+}) is constructed as follows:

1. $Cr_2O_7^{2-} \rightarrow 2Cr^{3+}$ The formula of the reactant and product are written (note that we write $2Cr^{3+}$ because there are two Cr species on the left hand side; ignore oxygen at this stage).
2. $Cr_2O_7^{2-} + 14H^+ \rightarrow 2Cr^{3+} + 7H_2O$ (We add 14 hydrogen ions to the left hand side because there are 7 oxygen atoms and these combine to form 7 molecules of water.)
3. The total charge on the left hand side is +12 ($-2 + 14 \times +1$) and the total charge on the right hand side is +6 ($2 \times +3$). To balance charge we add 6 electrons to the left hand side (we could take 6 electrons from the right hand side but it is easier to always add electrons).

The balanced half-equation is thus: $Cr_2O_7^{2-} + 14H^+ + 6e^- \rightarrow 2Cr^{3+} + 7H_2O$

We can see that this is a reduction reaction because electrons are gained and also by considering the oxidation number of the chromium species (this is the only species in the above equation that undergoes a change in oxidation number). The oxidation number of chromium in the dichromate ion is +6 (see earlier example) and is +3 in the chromium (III) ion. Because the oxidation number has decreased we know that the half-equation represents a reduction reaction.

4 A half-equation for a reduction process can be combined with one for an oxidation process to produce full equations for redox processes.

No electrons appear in *full* equations.

Consider the following two half-equations:

$MnO_4^- + 8H^+ + 5e^- \rightarrow Mn^{2+} + 4H_2O$ (a reduction half-equation)

$Fe^{2+} \rightarrow Fe^{3+} + e^-$ (an oxidation half-equation)

There are 5 electrons on the left hand side in the first equation and 1 electron on the right hand side in the second equation. By multiplying every species in the second equation by 5 we do not alter the meaning of the equation but we do allow the numbers of electrons on each side of the equation to be equal:

$$MnO_4^- + 8H^+ + 5e^- \rightarrow Mn^{2+} + 4H_2O$$

$$5Fe^{2+} \rightarrow 5Fe^{3+} + 5e-$$

Adding these equations together and cancelling any common terms allows the full equation to be derived:

$$5Fe^{2+} + MnO_4^- + 8H^+ + \cancel{5}e^- \rightarrow 5Fe^{3+} + \cancel{5}e^- + Mn^{2+} + 4H_2O$$

The final equation for the redox reaction between the manganate ion (MnO_4^-) and the iron (II) ion (Fe^{2+}) is therefore:

$$5Fe^{2+} + MnO_4^- + 8H^+ \rightarrow 5Fe^{3+} + Mn^{2+} + 4H_2O$$

Make sure that you:

Understand and can use the concept of oxidation number

Can balance half equations

Can combine half equations

Introduction to Oxidation and Reduction

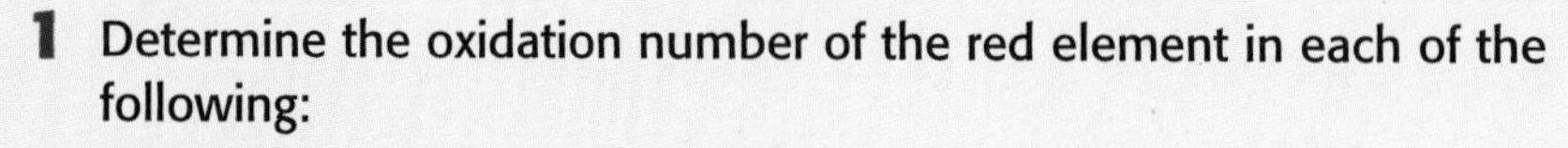

DAY 3

1 Determine the oxidation number of the red element in each of the following:

Learn the rules for calculating oxidation number

a) H_2SO_4 b) KIO_3 c) $S_2O_3^{2-}$

2

a) In each equation below, identify the species that has been oxidised:

Remember oxidation is an increase in oxidation number

i) $C + O_2 \rightarrow CO_2$ ii) $Zn + CuO \rightarrow ZnO + Cu$ iii) $2Cl^- \rightarrow Cl_2 + 2e^-$

b) In each equation below, identify the oxidising agent:

i) $PbO + Mg \rightarrow MgO + Pb$ ii) $Fe^{2+} + Ni \rightarrow Ni^{2+} + Fe$

3 Complete the following half-equations:

a) $MnO_4^- \rightarrow MnO_2$ b) $MnO_4^- \rightarrow Mn^{2+}$

c) $C_2O_4^{2-} \rightarrow CO_2$ d) $S_2O_3^{2-} \rightarrow S_4O_6^{2-}$

4 In order to determine the percentage of iron in a tablet, the tablet was dissolved in 25 cm^3 of dilute sulphuric acid. The solution was then titrated with potassium dichromate ($K_2Cr_2O_7$) solution. The dichromate ion ($Cr_2O_7^{2-}$) oxidised the iron (Fe^{2+}) to Fe^{3+}, itself being reduced to Cr^{3+}.

Remember that no electrons appear in a full equation

Deduce the reaction between Fe^{2+} and $Cr_2O_7^{2-}$ by constructing and combining half-equations.

Energetics

15 mins
Time Yourself

How much do you know?

1 a) Write an equation to represent each of the following enthalpy changes:
  i) Standard enthalpy of formation of ethanol (C_2H_5OH).
  ii) Standard enthalpy of combustion of carbon.
 b) What do you understand by the term standard conditions?

2 Using the enthalpy changes provided below:

$\Delta H_{c(C(s))} = -394\,kJ\ mol^{-1}$ $\Delta H_{c(H_2(g))} = -286\,kJ\,mol^{-1}$ $\Delta H_{c(C_2H_6(g))} = -1560$ $kJ\ mol^{-1}$ $\Delta H_{f(C_6H_{12}O_6(s))} = -1273\,kJ\,mol^{-1}$ $\Delta H_{f(C_2H_5OH(l))} = -277\,kJ\,mol^{-1}$

$\Delta H_{f(CO_2(g))} = -394\,kJ\,mol^{-1}$

 a) Calculate the enthalpy of formation of ethane:

$2C_{(s)} + 3H_{2(g)} \rightarrow C_2H_{6(g)}$

 b) Calculate the enthalpy change for the following reaction:

$C_6H_{12}O_{6(s)} \rightarrow 2C_2H_5OH_{(l)} + 2CO_{2(g)}$

3 Using the bond enthalpies below, calculate the enthalpy change for the following reaction:

$C_2H_{4(g)} + H_{2(g)} \rightarrow C_2H_{6(g)}$

C=C $612\,kJ\,mol^{-1}$ C–H $413\,kJ\,mol^{-1}$ H–H = $436\,kJ\,mol^{-1}$ C–C $347\,kJ\,mol^{-1}$

Answers

1a) i) $2C_{(s)} + 3H_{2(g)} + \frac{1}{2}O_{2(g)} \rightarrow C_2H_5OH_{(l)}$ ii) $C_{(s)} + O_{2(g)} \rightarrow CO_{2(g)}$ b) 298 K and 1 atm pressure
2a) $-86\,kJ\,mol^{-1}$ b) $-69\,kJ\,mol^{-1}$
3 $-125\,kJ\,mol^{-1}$

If you got them all right, skip to page 36

Spend no more than 30 mins on this topic

Learn the key facts

1 The enthalpy change (ΔH) of a reaction is defined as the heat change measured under constant pressure.

The standard enthalpy of formation (ΔH_f^{θ}) of a compound is the enthalpy change when one mole of a substance is formed from its constituent elements under standard conditions.

The standard enthalpy of combustion (ΔH_c^{θ}) is the enthalpy change when one mole of a substance is completely burnt under standard conditions.

The standard enthalpy of neutralisation (ΔH_{neut}^{θ}) is the enthalpy change when one mole of H^+ ions (from an acid) combines with one mole of OH^- ions (from an alkali) under standard conditions.

Standard conditions are 298 K and 1 atmosphere pressure.

2 Hess's law states that the enthalpy change for a reaction is independent of the route taken, depending only on the initial and final states. Knowing the above definitions and setting out calculations as overleaf enables us to solve Hess's Law problems.

Calculate the enthalpy of formation of ethanol given the following enthalpies of combustion:

$C = -393\,kJ\,mol^{-1}$ $H_2 = -286\,kJ\,mol^{-1}$ $C_2H_5OH = -1367\,kJ\,mol^{-1}$

- The equation for the reaction for which you are trying to determine the enthalpy change should be written across the top:
  $2C_{(s)} + 3H_{2(g)} + \frac{1}{2}O_{2(g)} \rightarrow C_2H_5OH_{(l)}$
- Writing equations to represent the combustion of C, H_2 and C_2H_5OH reveals that the products are common, i.e. water and carbon dioxide. Thus these products are written below the equation (it is not always necessary to balance oxygen).
- Arrows are drawn to represent the enthalpy changes. The arrows should be labelled with the enthalpy change they represent.

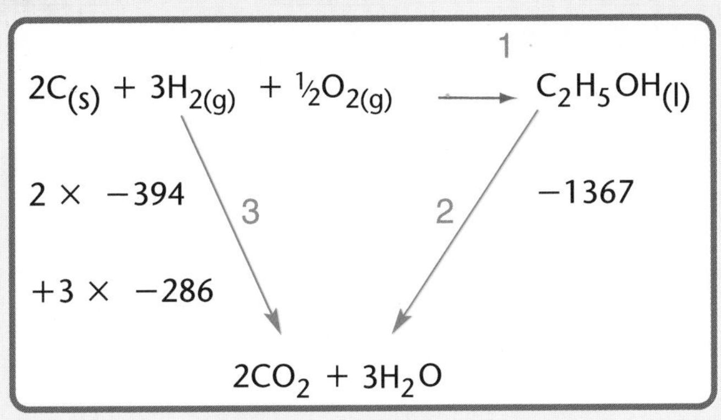

- Applying Hess's Law allows us to state that the enthalpy change labelled 1 is equal to 3 − 2.
- Substituting in the figures allows us to determine the enthalpy of formation of ethanol, i.e. of $-279\,kJ\,mol^{-1}$.

3 This approach can be used to solve problems that involve bond enthalpies. It is important to appreciate that bond formation is an exothermic process and bond breaking is an endothermic process. Often questions do not distinguish whether the enthalpy value given is for bond forming or breaking but the values are identical; it is only the preceding sign that changes.

Bond enthalpy values are defined as the energy required to break one mole of a certain bond (e.g. the first C–H bond in methane). Average bond enthalpy values correspond to the average bond enthalpy for one type of bond (e.g. the average of the four C–H bonds in methane).

Calculate the enthalpy change for the following reaction given the following bond enthalpies:

$CH_4 + 2O_2 \rightarrow CO_2 + 2H_2O$

C–H $413\,kJ\,mol^{-1}$; O=O $498\,kJ\,mol^{-1}$; C=O $805\,kJ\,mol^{-1}$; O–H $464\,kJ\,mol^{-1}$.

Drawing the structural formulae of the above compounds and considering the bonds broken/formed shows that the bonds broken are 4 × C–H and 2 × O=O. Bond breaking is an endothermic process and so the total energy input is $2648\,kJ\,mol^{-1}$ [(4 × 413) + (2 × 498)]. The bonds formed are 2 × C=O and 4 × O–H. Bond formation is an exothermic process and so the total energy released is $-3466\,kJ\,mol^{-1}$. The net energy change is $2648 - 3466 = -818\,kJ\,mol^{-1}$.

Make sure that you:

Know the four enthalpy definitions

Understand and can apply Hess's Law

Can use bond enthalpies to calculate energy changes

Have you improved?

1 Write an equation to represent each of the following standard enthalpies:
a) formation of glucose ($C_6H_{12}O_6$)
b) atomisation of carbon
c) neutralisation (for the reaction between NaOH and H_2SO_4)

Learn these three definitions

2 a) The equation for the reduction of ethanal to ethanol is:
$CH_3CHO + H_2 \rightarrow CH_3CH_2OH$

Using the enthalpy values below, determine the enthalpy change for this reaction:

ΔH_c^ϑ (CH_3CHO) $= -1167\,kJ\,mol^{-1}$

ΔH_c^ϑ (H_2) $= -286\,kJ\,mol^{-1}$

ΔH_c^ϑ (CH_3CH_2OH) $= -1368\,kJ\,mol^{-1}$

Write out equations for the enthalpy changes given and construct an energy cycle

3 For the following reaction:

$C_5H_{12} + 8O_2 \rightarrow 5CO_2 + 6H_2O$

Determine the enthalpy change of this reaction given the following bond energy data:
C–C 347 $kJ\,mol^{-1}$
C–H 413 $kJ\,mol^{-1}$
O=O 498 $kJ\,mol^{-1}$
C=O 736 $kJ\,mol^{-1}$
H–O 464 $kJ\,mol^{-1}$

Write structural formulae to see which bonds are broken/formed

Kinetics

How much do you know?

1 a) List two factors (other than concentration) that affect rates of reaction.
b) In terms of particles, explain how doubling the concentration of acid increases the rate of reaction with marble chips.

2 a) Sketch the energy distribution for a gaseous sample at two different temperatures. State which is the higher temperature.
b) Explain how the presence of a catalyst increases the rate of the reaction.

3 Explain the following terms:
a) thermodynamic stability
b) kinetic stability.

Answers

1a) Any two from temperature, surface area, presence of a catalyst, pressure (for reactions involving gases) b) Doubling the concentration of acid doubles the number of acid particles (H^+) present. This doubles the chance of a collision with the marble chip and hence the rate of the reaction increases.
2a) See the diagram on page 39. b) Provides an alternative pathway of lower activation energy, hence more particles have the required energy to react.
3a) Products are lower in energy than the reactants. b) The reaction has a high activation energy.

If you got them all right, skip to page 41

Kinetics

Learn the key facts

1 The following factors control the rates of chemical reactions:

- concentration
- temperature
- pressure (for reactions involving gases)
- surface area
- catalysts

Collision theory helps to explain the effect of changing any of these variables on the rate of reaction. In order for a chemical reaction to occur it is necessary for the reactant molecules to 'collide' with each other. Also, these particles must collide with enough energy for a reaction to occur. This energy is known as the activation energy.

Consider the reaction between calcium carbonate (marble chips) and hydrochloric acid:

$$CaCO_{3(s)} + 2HCl_{(aq)} \rightarrow CaCl_{2(aq)} + H_2O_{(l)} + CO_{2(g)}$$

The rate of reaction can be measured by recording the volume of gas produced per unit time. By increasing the concentration of acid, gas is produced more quickly, indicating that the rate of reaction has increased. Similarly, using powdered marble also causes an increase in reaction rate.

Collision theory tells us that in order for the reaction to occur the acid particles must collide with the marble chips. At a higher concentration of acid there are more acid particles per unit volume and therefore a higher chance that a collision will occur. The greater number of collisions means there is a higher chance that more collisions will have the required activation energy and so the rate of reaction increases.

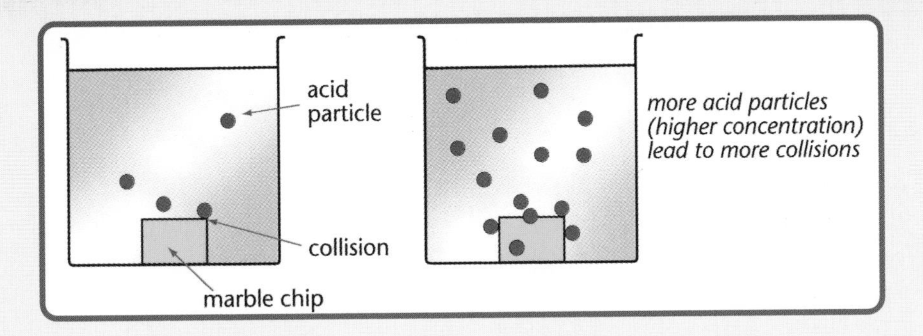

By increasing the surface area of the marble we have created more places where a collision can occur. The greater number of collisions means that the number of 'successful' collisions increases (i.e. collisions with the activation energy); hence the reaction rate also increases.

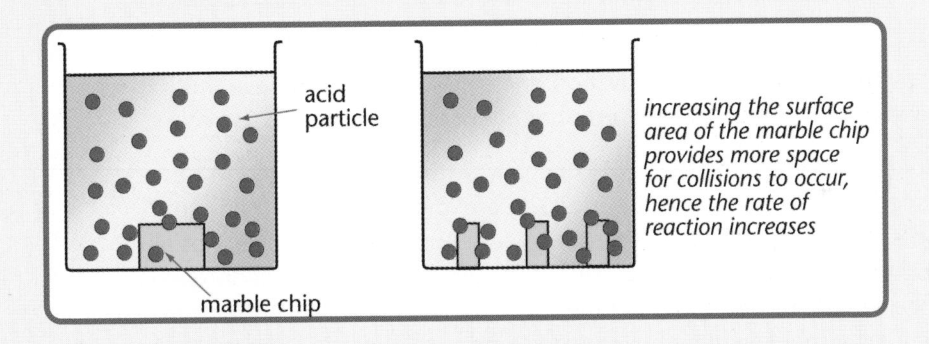

2 The Maxwell-Boltzmann distribution is a plot of the energies of all the particles in a given sample.

The first curve shows the energy distribution at an initial temperature T_1; the second curve shows the energy distribution at a higher temperature T_2. The line labelled E_A represents the activation energy. This is the energy that the particles must possess in order for a successful collision, i.e. for a reaction to occur. The shaded area under the curve represents the number of particles with a higher energy than the activation energy. The area shaded under curve T_2 is greater than the area shaded under curve T_1 hence at a higher temperature there is a greater rate of reaction.

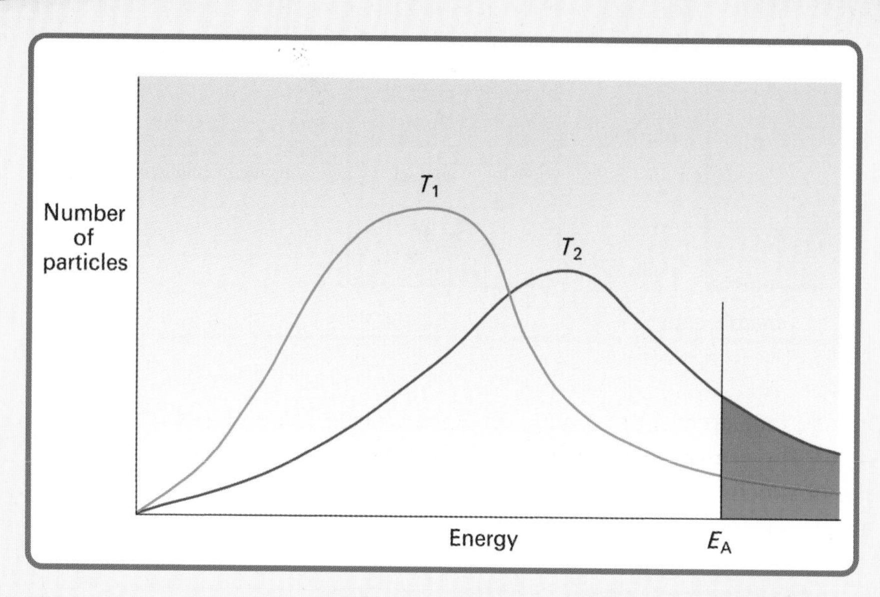

The presence of a catalyst increases the rate of reaction by providing an alternative pathway of lower activation energy. More particles have this activation energy and hence the number of successful collisions increases.

3 Reactions that are exothermic (ΔH is negative) might be expected to be spontaneous. This is because the products are lower in energy than the reactants, i.e. the products are thermodynamically stable. However, not all exothermic reactions are spontaneous (for example the combustion of methane requires a flame to start the reaction). This is because the activation energy for the reaction is high. Reactions with high activation energy are said to be kinetically stable.

Make sure that you:

Know the factors that control reaction rates

Understand and can apply collision theory

Understand the concepts of thermodynamic and kinetic stability

Have you improved?

1 The equation for the reaction of marble chips (calcium carbonate) with dilute hydrochloric acid is:

$$CaCO_{3(s)} + 2HCl_{(aq)} \rightarrow CaCl_{2(aq)} + H_2O_{(l)} + CO_{2(g)}$$

a) (i) What will be the effect on the rate of the reaction of increasing the temperature of the acid?

The acid particles move more quickly at a higher temperature

(ii) Explain your answer.

Remember activation energy

b) (i) What will be the effect of using powdered calcium carbonate instead of marble chips on the rate of reaction?

How has the surface area changed?

(ii) Explain your answer.

2 a) Using the same set of axes, draw two curves to show the distribution of energies in the same sample of gas molecules at two temperatures T_1 and T_2, where T_2 is the higher temperature. Label the axes and curves.

Learn this!

b) What does the area under each curve represent?

c) With reference to your diagram in a) explain why an increase in temperature increases the rate of a gaseous reaction.

Consider your answer to b)

d) Explain why the presence of a catalyst increases the rate of a reaction.

Consider activation energy

3 Although the standard enthalpy change for the oxidation of methanol to methanal is negative, methanol is stable at room temperature. How do you account for this?

Consider thermodynamic and kinetic stability

Equilibria

How much do you know?

1 What do you understand by the term dynamic equilibrium?

2 For the following equilibrium:

$$N_{2(g)} + 3H_{2(g)} \rightleftharpoons 2NH_{3(g)} \quad \Delta H = -92\,kJ\,mol^{-1}$$

State the effect of:

a) an increase in the total pressure
b) an increase in nitrogen concentration
c) a decrease in temperature.

3 In the industrial production of ammonia a compromise temperature is used. What do you understand by the term compromise temperature?

Answers

1 An equilibrium reaction where the rate of the forward reaction is equal to the rate of the reverse reaction. **2a)** Equilibrium moves to the RHS (favouring the reaction that produces the fewer number of moles of gas), K_c remains unchanged. b) Equilibrium moves to the RHS, K_c remains unchanged. c) The reaction favours the exothermic reaction, i.e. will move to the RHS, K_c will increase.
3 In the manufacture of ammonia a low temperature favours the formation of ammonia. However, if the temperature is too low then the rate of the reaction is too slow, so a 'medium' temperature is used. This temperature is called a compromise temperature.

If you got them all right, skip to page 45

Equilibria

Learn the key facts

1 Many chemical reactions are reversible, i.e. can go forwards and backwards. Reversible reactions are represented by the $\rightleftharpoons$ symbol instead of an arrow. Reversible reactions can reach equilibrium. At equilibrium the concentrations of reactants and products remain constant because the rate of the forward reaction is equal to the rate of the reverse reaction.

2 Le Chatelier's principle states that when a change is imposed upon a system at equilibrium the system acts to oppose that change. Consider an increase in the concentration of ethanoic acid in the following equilibrium:

$$CH_3COOH_{(aq)} + C_2H_5OH_{(aq)} \rightleftharpoons CH_3COOC_2H_{5(aq)} + H_2O_{(aq)}$$

The system wants to remove the extra acid and so the equilibrium position moves to the right. For gaseous equilibrium (see below) the effect of pressure is the same as concentration.

The effect of temperature depends upon whether the reaction is exothermic or endothermic. If the forward reaction is exothermic (e.g. in the Haber process) then an increase in temperature favours the endothermic path (i.e. the reverse reaction). However, an increase in temperature increases the kinetic energy of the gas molecules, hence there will be more successful collisions. As a result a compromise temperature is used in the manufacture of ammonia.

The presence of a catalyst in equilibrium reactions (e.g. Fe in the Haber process) serves only to reduce the time taken for equilibrium to be reached. The catalyst does not increase the yield.

The manufacture of ammonia (Haber process) is a good example of an industrial equilibrium process.

$$N_2 + 3H_2 \rightleftharpoons 2NH_3$$

This reaction is carried out at 200 atmospheres pressure and 450 C in the presence of a finely divided iron catalyst. The presence of a catalyst is simply to reduce the time taken for the reaction to reach equilibrium.

The high pressure causes the equilibrium to move to the right-hand side (according to Le Chatelier's principle) because a high pressure favours the side of the equation with the least number of moles of gas. A higher pressure would increase the yield of ammonia but the increased cost makes a higher pressure uneconomical. Also, a higher pressure is very dangerous.

According to Le Chatelier's principle, a low temperature favours the formation of ammonia (low temperatures favour the endothermic pathway). However, a low temperature causes a slow rate of reaction (see page 38). As a result, a 'compromise' temperature is used that optimises the yield and the rate of formation of ammonia.

Make sure that you:

Understand the concept of dynamic equilibrium

Can apply Le Chatelier's principle

Understand the selection of temperature in the manufacture of ammonia

Have you improved?

1 What can you infer about the concentrations of gases in the manufacture of ammonia at equilibrium?

Think about the rate of the reaction

2 Consider the following equilibrium:

$$2CrO_4^{2-}{}_{(aq)} + 2H^+{}_{(aq)} \rightleftharpoons Cr_2O_7^{2-} + H_2O$$

yellow orange

a) If the concentration of acid is increased, what colour would you expect your solution to be?
b) Explain your answer to part i).
c) Given that the forward reaction is exothermic, explain the effect on the equilibrium of decreasing the temperature.

3 The industrial manufacture of ammonia is carried out at approximately 450°C.

The forward reaction is exothermic

$$N_{2(g)} + 3H_{2(g)} \rightleftharpoons 2NH_{3(g)} \qquad \Delta H = -92\,kJ\,mol^{-1}$$

a) What will be the effect on the yield of ammonia if the temperature is increased?
b) With reference to your answer in (i) explain why a lower temperature is not used.

Consider kinetics

Groups 1 and 2

How much do you know?

1 How does atomic radius vary down groups 1 and 2?

2 How does the strength of metallic bonding vary down groups 1 and 2?

3 What colour does sodium impart to a flame?

4 Name the products formed when potassium reacts with water.

5 Complete the equation: $MgO + 2HCl \rightarrow$

6 Name the compound Na_2O_2.

7 What type of bonding is present in $MgCl_2$?

8 In what way is $Be(OH)_2$ unusual for group 2?

9 What is the trend in solubility of group 2 hydroxides?

10 Name the products formed when KNO_3 decomposes.

11 Complete the equation: $CaCO_3 + 2HCl \rightarrow$

12 What is lime water?

Answers

1 increases **2** becomes weaker **3** yellow **4** potassium hydroxide and hydrogen **5** $\rightarrow MgCl_2 + H_2O$ **6** sodium peroxide **7** ionic **8** it is amphoteric **9** become more soluble **10** potassium nitrite and oxygen **11** $\rightarrow CaCl_2 + CO_2 + H_2O$ **12** aqueous $Ca(OH)_2$

If you got them all right, skip to page 49

Groups 1 and 2

Learn the key facts

1 Groups 1 and 2 consist of the relatively reactive metals Li–Fr and Be–Ra. Descending the group, successive electron shells are filled so atomic radii and shielding increase, reducing nuclear attraction for the outer electrons. Thus, ionisation energies decrease and group 1 and 2 metals become more reactive, forming M^+ and M^{2+} ions respectively (oxidation states +1 and +2).

2 Descending the group, cationic radii increase, reducing surface charge density, and hence metallic bond strength decreases. Therefore, melting points and physical strength decrease.

3 Group 1 and 2 metals can be detected by flame tests. A platinum wire is dipped into a sample of the solid compound or solution and held in a strong flame. The emission spectra of the element is seen as a coloured flame.

Metal	Li	Na	K	Ca	Sr	Ba
Colour	scarlet	yellow	lilac	brick red	crimson	apple green

4 Group 1 and 2 metals react with H_2O to form alkaline hydroxides and H_2 gas (except Mg which reacts with steam to form an insoluble oxide).

$Li + H_2O \rightarrow LiOH + \frac{1}{2}H_2$ and $Ca + 2H_2O \rightarrow Ca(OH)_2 + H_2$

$Mg + H_2O \rightarrow MgO + H_2$

5 Group 1 and 2 metals react with O_2 to form ionic, basic oxides (except Be). These oxides dissolve in water (except MgO and BeO) to form alkaline hydroxide solutions and all react with acids to form salts.

$4Na + O_2 \rightarrow 2Na_2O$ and $2Sr + O_2 \rightarrow 2SrO$

$K_2O + H_2O \rightarrow 2KOH$ and $MgO + 2HCl \rightarrow MgCl_2 + H_2O$

6 Heavier group 1 metals also form peroxides, containing O_2^{2-} anions (O in the −1 state) and super oxides containing O_2^- anions (O in the $-\frac{1}{2}$ state).

$2K + O_2 \rightarrow K_2O_2$ (potassium peroxide)

$Rb + O_2 \rightarrow RbO_2$ (rubidium superoxide)

Groups 1 and 2

7 Group 2 metals form ionic chlorides with Cl_2 gas: $Mg + Cl_2 \rightarrow MgCl_2$

8 Be shows some atypical properties. Its hydroxide is amphoteric – it reacts with acids *and* bases. In the solid state its chloride is covalent due to the powerful polarising effect of the small Be^{2+} ion which distorts the anion.

$Be(OH)_2 + 2OH^- \rightarrow Be(OH)_4^{2-}$ 4-coordinate due to small size of Be^{2+} ion.

$Be(OH)_2 + 2HCl \rightarrow BeCl_2$

9 Descending group 2 hydroxides become more soluble; $Mg(OH)_2$ is only sparingly soluble. Sulphates become less soluble; $-BaSO_4$ is very insoluble and the formation of this dense, white ppt. is a chemical test for sulphate ions.

10 Group 1 and 2 nitrates undergo thermal decomposition. Nitrate anions are distorted and destabilised by polarising cations so as the cation charge density decreases down the group nitrates become more stable. Similarly group 2 nitrates – which form the oxide, NO_2 and O_2 – are less stable than those of group 1 – which form the nitrite and O_2 (except Li which decomposes like a group 2 metal due to its unusually high charge density).

$NaNO_3 \rightarrow NaNO_2 + \frac{1}{2}O_2$

$2LiNO_3 \rightarrow Li_2O + 2NO_2 + \frac{1}{2}O_2$ and $Ca(NO_3)_2 \rightarrow CaO + 2NO_2 + \frac{1}{2}O_2$

11 Group 2 carbonates thermally decompose to form oxides and CO_2 gas. Magnesium and calcium carbonate are insoluble and occur naturally as minerals such as dolomite and lime stone respectively. Both react readily with strong acids to produce salts, CO_2 and H_2O.

$CaCO_3 \rightarrow CaO + CO_2$ and $MgCO_3 + 2HCl \rightarrow MgCl_2 + CO_2 + H_2O$

12 CaO is mixed with H_2O to produce $Ca(OH)_2$ ('slaked lime'), used to make cement, plaster and to increase soil pH. Dilute aqueous $Ca(OH)_2$ (lime water) is used to test for CO_2 gas, forming a ppt. of $CaCO_3$, which dissolves in excess CO_2 to form $Ca(HCO_3)_2$ (also present in hard water).

$Ca(OH)_2 + CO_2 \rightarrow CaCO_3 + H_2O$ then $CaCO_3 + H_2O + CO_2 \rightarrow Ca(HCO_3)_2$

Make sure you are familiar with the following key terms:

- *trends in atomic radii, ionisation energy, reactivity and melting point*
- *flame colours*
- *reactions with water, oxygen and chlorine*
- *basic hydroxides and oxides*
- *solubility of hydroxides and sulphates*
- *thermal decomposition of nitrates and carbonates*
- *uses of $CaCO_3$ and $Ca(OH)_2$*

Have you improved?

1 a) Explain why caesium melts at a lower temperature than lithium.
b) Explain why caesium reacts violently with water to form an aqueous hydroxide but lithium does so much more slowly.
c) Write an equation for the reaction between magnesium and water including state symbols. Describe the conditions required and explain why the Mg compound you have stated is formed.

Metallic bond strength

Ionisation energy

Forms MgO

2 a) Write equations for the formation of calcium oxide from its elements and its subsequent reaction with nitric acid. What property does the oxide show in the second reaction?
b) Describe and explain how a solution of calcium ions and another of barium ions could be distinguished using both flame tests and the addition of sulphuric acid.
c) Describe, giving equations, how a solution of $Ca(OH)_2$ can be used to test for the presence of CO_2 gas. Describe also what occurs when CO_2 is present in excess.

CaO

Flame colours and solubility

Formation of $CaCO_3$ ppt

3 a) Describe why calcium nitrate decomposes at a lower temperature than potassium nitrate. Write an equation for the decomposition of calcium nitrate.
b) State a naturally occurring source of calcium carbonate and explain, giving an equation, why it is particularly susceptible to weathering by acid rain.
c) Beryllium chloride is unusual in that it has covalent bonding. Explain why this is so.

Cation charge density

Carbonate + acid gives…?

Be^{2+} very small

Group 7

How much do you know?

1 What is the trend in boiling point of the halogens down group 7?

2 Complete the equation
$Cl_2 + 2I^- \rightarrow ?$

3 Which is the stronger reducing agent: Br^- or I^-?

4 What is seen when $AgNO_3(aq)$ is added to a solution of I^- ions?

5 What electrolyte is used for the production of chlorine?

6 Complete the equation
$2I^- + ClO^- + H_2O \rightarrow ?$

7 What kind of acid is HCl?

Answers

1 increases **2** $\rightarrow 2Cl^- + I_2$ **3** I^- **4** yellow ppt **5** molten NaCl
6 $\rightarrow I_2 + Cl^- + 2OH^-$ **7** strong

If you got them all right, skip to page 53

Group 7

Spend no more than 30 mins on this topic

Learn the key facts

1 Descending group 7 atomic radii and shielding increase, leading to stronger van der Waal's intermolecular forces and hence boiling points increase. F_2 and Cl_2 are pale yellow and green gases respectively. Br_2 is a volatile, red-brown liquid and I_2 is a black solid.

2 Oxidising power and electronegativity decrease down the group as increasing atomic radii and shielding reduce nuclear attraction for outer electrons. E.g. displacement reactions: halogens are oxidised from solutions of their ions by a 'higher' halogen (this can be used as a chemical test).

	Cl_2	Br_2	I_2
Cl^-		no change	no change
Br^-	red brown solution of Br_2		no change
I^-	black ppt/yellow soln. of I_2	black ppt/yellow soln. of I_2	

This reaction is the basis of the extraction of bromine from sea water.

E.g. $2NaBr(aq) + Cl_2(g) \rightarrow 2NaCl(aq) + Br_2(aq)$

Other tests for the elements include:

Cl_2: turns damp, blue litmus paper red and then bleaches.

I_2: gives a blue-black colour with starch and is liberated by ClO^- ions.

3 Halide ions may be identified using aqueous $AgNO_3$ to produce coloured silver halide precipitates ($Ag^+(aq) + X^-(aq) \rightarrow AgX(s)$) distinguished by their solubility in aqueous NH_3.

	$+AgNO_3(aq)$	$+$dilute $NH_3(aq)$	$+$conc. $NH_3(aq)$
Cl^-	white AgCl ppt	soluble	soluble
Br^-	cream AgBr ppt	insoluble	soluble
I^-	yellow AgI ppt	insoluble	insoluble

4 Halide ions become increasingly powerful reducing agents as ionic radii increases down the group and nuclear attraction for outer electrons decreases. E.g. the reaction between solid halide salts and concentrated H_2SO_4.

$NaCl + H_2SO_4 \rightarrow NaHSO_4 + HCl$ (Cl^- not strong enough to reduce H_2SO_4)

$NaBr + H_2SO_4 \rightarrow NaHSO_4 + HBr$ (Br^- then reduces H_2SO_4)

$2HBr + H_2SO_4 \rightarrow SO_2 + Br_2 + 2H_2O$ (Br from -1 to 0, S from $+6$ to $+4$)

$8HI + H_2SO_4 \rightarrow 4I_2 + H_2S + H_2O$ (S from $+6$ to -2)

5 Chlorine is produced by the electrolysis of molten NaCl.

Anode reaction: $2Cl^- \rightarrow Cl_2 + 2e^-$ Cathode reaction: $Na^+ + e^- \rightarrow Na$

Chlorine is used to make HCl, antiseptics and bleaches and for anti-bacterial treatment of drinking water. It is also used to produce sodium chlorate(I), NaClO, by reaction with NaOH from the electrolysis of aqueous NaCl.

Anode reaction: $2Cl^- \rightarrow Cl_2 + 2e^-$ Cathode reaction: $2H^+ + 2e^- \rightarrow H_2$

H^+ from H_2O is liberated in preference to Na^+ leaving aqueous NaOH.

Cl_2 undergoes disproportionation with NaOH; Cl is reduced *and* oxidised. NaClO is used as bleach and to estimate I_2. If NaClO is heated, further disproportionation forms sodium chlorate(V), $NaClO_3$ (used as weed killer).

$$2NaOH + \underset{\underline{0}}{Cl_2} \rightarrow Na\underset{\underline{-1}}{Cl} + Na\underset{\underline{+1}}{Cl}O + H_2O \text{ and } 3Na\underset{\underline{+1}}{Cl}O \rightarrow 2Na\underset{\underline{-1}}{Cl} + Na\underset{\underline{+5}}{Cl}O_3$$

6 Oxidants (e.g. NaClO) can be estimated by reaction with I^- ions, or the I_2 liberated can be titrated with sodium thiosulphate $Na_2S_2O_3$.

$$2I^- + ClO^- + H_2O \rightarrow I_2 + Cl^- + 2OH^- \text{ and } I_2 + 2S_2O_3^{2-} \rightarrow 2I^- + S_4O_6^{2-}$$

7 Hydrogen halides, HX, are covalent, molecular gases. HF has an elevated boiling point due to hydrogen bonding. Most are strong acids, fully dissociating in water to form hydrated hydrogen ions. HF is a weak acid because the strong H–F bond does not easily break. The presence of hydrogen bonding also suppresses dissociation.

$HCl + H_2O \rightarrow H_3O^+ + Cl^-$

Make sure you are familiar with the following key terms:

- *trends in boiling point, electronegativity, oxidising power of halogens*
- *physical appearance of halogens*
- *displacement*
- *tests for halogens*
- *test for halide ions*
- *reducing power of halides*
- *halides + conc. H_2SO_4*
- *production of Cl_2, ClO^- and ClO_3^- by electrolysis and their uses*
- *disproportion*
- *hydrogen halides as acids*

Have you improved?

1 a) Explain why iodine is solid whereas fluorine is a gas.

b) Explain why chlorine is a stronger oxidising agent than bromine and how it is used for the extraction of bromine.

c) Give an equation for the formation of sodium chlorate (I) and explain what type of reaction it is. Give different uses of chlorine and sodium chlorate (I).

2 a) When solid NaCl is treated with conc. H_2SO_4, white misty fumes are observed. Write an equation for this reaction, identify the gas and suggest how it could be tested for.

b) Write equations to show how sodium bromide reacts with conc H_2SO_4 and explain why it behaves differently to NaCl.

3 a) Describe how solutions of chloride and bromide ions may be distinguished other than by displacement.

b) Describe briefly how oxidising agents may be estimated using solutions of KI and sodium thiosulphate, $Na_2S_2O_3$.

c) Write an equation to show that HBr is a strong acid and explain why HF is a weak acid.

Intermolecular forces

Cl smaller than Br

Cl_2 + NaOH

HCl produced

Redox reaction

$AgNO_3$

I_2 liberated

Dissociation in H_2O

15 mins
Time Yourself

Industrial Inorganic Chemistry

How much do you know?

1 What catalyst is used in the manufacture of sulphuric acid?

2 What is the name of the process by which NH_3 is produced?

3 Give an equation for the oxidation of NO to NO_2.

4 What three elements, essential for plant growth, are commonly supplied by fertilisers?

5 What is the main reducing agent in the blast furnace used for the production of iron?

6 By what process is Al metal extracted from purified Al_2O_3?

7 Which compound is reduced to produce Ti metal?

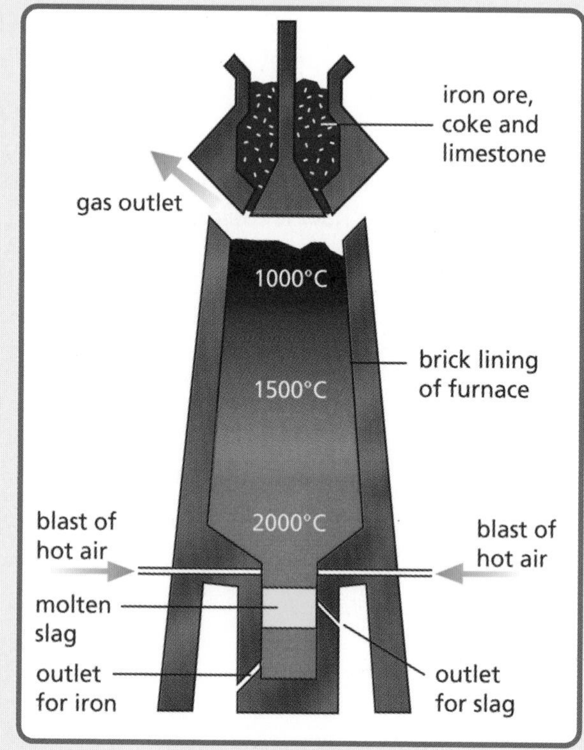

Answers

1 V_2O_5
2 Haber
3 $NO + \frac{1}{2}O_2 \rightarrow NO_2$ or $2NO + O_2 \rightarrow 2NO_2$
4 N, P and K
5 CO (C)
6 electrolysis
7 $TiCl_4$

If you got them all right, skip to page 59

Industrial Inorganic Chemistry

Spend no more than 30 mins on this topic

Learn the key facts

1 Sulphuric acid is produced by the contact process.

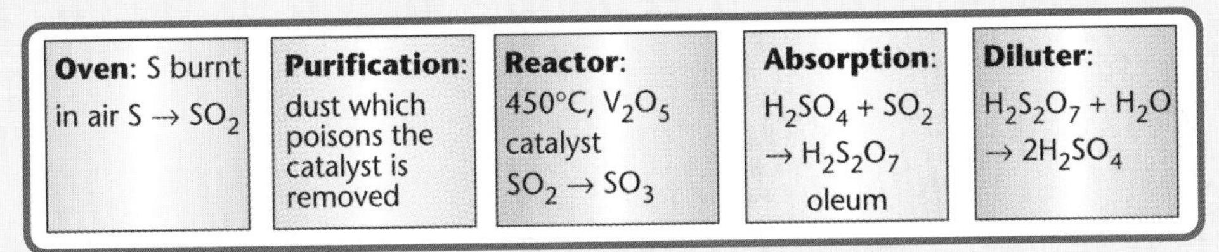

$2S + O_2 \rightarrow 2SO_2$ and then $2SO_2 + O_2 \rightarrow 2SO_3$ $\Delta H = -197\,kJ\,mol^{-1}$

- The exothermic second step controls the conditions. Low temperatures give high yields but a slow rate. Rates are improved by higher temperatures (yield still 97% at 450 °C), high pressures of SO_2 and O_2 and a catalyst.
- Heat generated by the second step is removed by heat exchangers to prevent overheating and reduced yields and is used to heat the incoming gases, reducing the need for additional, external heating.
- H_2SO_4 is used to make fertilisers, dyes, polymers, detergents and drugs.

2 Ammonia is manufactured by the Haber process.

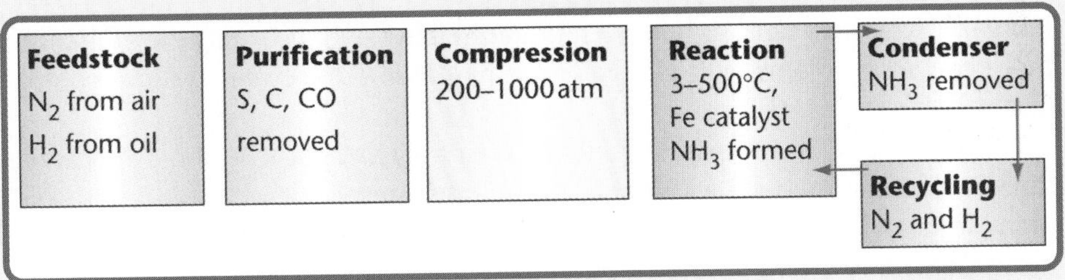

$N_2 + 3H_2 \rightarrow 2NH_3$ $\Delta H = -92\,kJ\,mol^{-1}$

- Temperature is a compromise between acceptable rate (high T) and yield (low T). Yield is improved by recycling unreacted N_2 and H_2.
- NH_3 is used to manufacture fertilisers, nitric acid, and nylon.

3 Nitric acid is manufactured by the Ostwald process.

Feedstock:	Reactor:	Cooler:	Distillation:	Absorption
NH_3 from Haber process O_2 from air	900°C Pt/Rh catalyst, NH_3 to NO	air oxidises NO to NO_2	pure HNO_3 formed	$NO_2 + H_2O \rightarrow HNO_3$

$$4NH_3 + 5O_2 \rightarrow 4NO + 6H_2O \quad \Delta H = -950 \text{ kJ mol}^{-1}$$

$$2NO + O_2 \rightarrow 2NO_2 \quad \Delta H = -114 \text{ kJ mol}^{-1}$$

$$3NO_2 + H_2O \rightarrow 2HNO_3 + NO \quad \Delta H = -117 \text{ kJ mol}^{-1}$$

- High temperatures are necessary for stage 1, despite being exothermic, to overcome a large activation energy and ensure a viable rate.
- Nitric acid is used to manufacture dyes, explosives and fertilisers.

4 Commercial fertilisers are widely used to increase the nitrogen content of soil essential for the production of proteins and chlorophyll.

- Ammonium salts and nitrates (e.g. NH_4NO_3) are widely used because of their high N content. However, ammonium cations are acidic, decreasing the pH of the soil, which can adversely affect plant growth. They can also reduce the calcium content of the soil.
- Inorganic fertilisers tend to release nitrogen relatively quickly into the soil and so there can be problems with rainwater washing it into the water system, contributing to eutrophication.
- Urea, $(NH_2)_2C{=}O$, is an organic fertiliser with a high N content and the advantage that it releases N at a slow rate and doesn't reduce the soil pH.
- Many fertilisers also provide phosphorus and potassium for plant growth and these are classified according to their NPK content.

5 Iron is produced by reduction of iron(III) oxide in a blast furnace.

- Coke (C), iron ore (Fe_2O_3) and limestone ($CaCO_3$), are heated in the furnace.
- Coke burns, heating the reaction and forms CO_2 and CO which reduces the oxide to iron metal.
- Silica impurities are removed by conversion to calcium silicate or slag.

$C + O_2 \rightarrow CO_2$	Exothermic
$CO_2 + C \rightarrow 2CO$	
$3CO + Fe_2O_3 \rightarrow 2Fe + 3CO_2$	Reduction
$CaO + SiO_2 \rightarrow CaSiO_3$	Slag

- The molten 'pig' iron is tapped off from the furnace and carbon impurities are removed as CO_2 by blowing O_2 gas through the melt. Carbon content affects the physical properties of the iron: high carbon yields hard, brittle cast iron; low carbon gives soft, malleable mild steel.
- Iron is widely used because it is abundant in high grade ores, cheap to produce and strong. It is used to make cars, buildings, bridges, etc.
- The main disadvantage of iron is that it readily corrodes (rusts) in atmospheric conditions, i.e. H_2O and O_2. Iron is oxidised to iron (II) hydroxide, then to iron (III) hydroxide, which then forms iron (III) oxide.

  $Fe + H_2O + \frac{1}{2}O_2 \rightarrow Fe(OH)_2 \rightarrow Fe(OH)_3 \rightarrow Fe_2O_3$

  This problem can be reduce by alloying iron with other metals to produce steels resistant to rusting.

6 Aluminium is produced by electrolysis of Al_2O_3 extracted from bauxite.

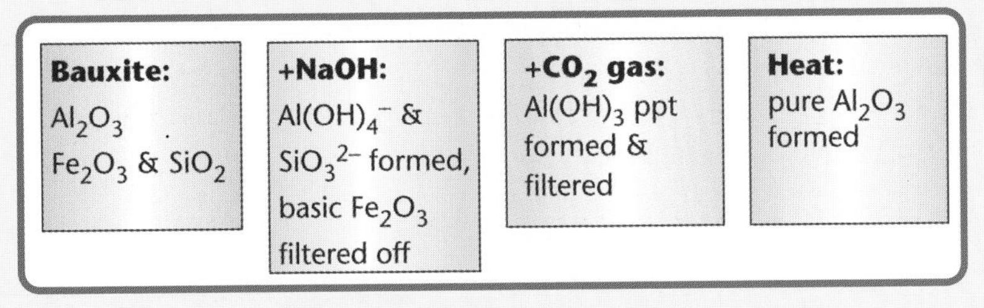

Bauxite:	**+NaOH:**	**+CO_2 gas:**	**Heat:**
Al_2O_3, Fe_2O_3 & SiO_2	$Al(OH)_4^-$ & SiO_3^{2-} formed, basic Fe_2O_3 filtered off	$Al(OH)_3$ ppt formed & filtered	pure Al_2O_3 formed

$Al_2O_3 + 2OH^- + 3H_2O \rightarrow 2Al(OH)_4^-$ (Al species acidic)

$SiO_2 + 2OH^- \rightarrow SiO_3^{2-}$ (SiO_2 acidic)

$2Al(OH)_4^- + CO_2 \rightarrow 2AlOH_3 + CO_3^{2-} + H_2O$ (Al species basic)

- Al_2O_3 is mixed with molten cryolite, Na_3AlF_6, depressing its melting point to 850 °C. The mixture is then electrolysed:

Cathode reaction: $Al^{3+} + 3e^- \rightarrow Al$ Anode reaction: $2O^{2-} \rightarrow O_2 + 4e^-$

- Electrolysis is used because Al is high in the electrochemical series and forms strong bonds with O so chemical reductants (e.g. C) are not powerful enough to liberate the metal.
- Electrolysis requires a lot of electrical energy, so plants are often sited near cheap electricity sources, e.g. hydroelectric power.
- Aluminium is light, strong and resistant to corrosion because it forms an impervious oxide layer. It is used for window frames, kitchen utensils, cars and aircraft construction.

7 Titanium is extracted by reduction of the $TiCl_4$ by active metals. Direct reduction of the oxide is highly endothermic and so uneconomical.

- $TiCl_4$ is produced from rutile, TiO_2 and Cl_2, followed by distillation. $TiCl_4$ is reduced by molten sodium or magnesium under an inert argon atmosphere to prevent potentially dangerous combustion of the metals.

  $TiO_2 + 2Cl_2 + 2C \rightarrow TiCl_4 + 2CO$ and $TiCl_4 + 4Na \rightarrow Ti + 4NaCl$

- This process is expensive due to the materials used and because it is a batch process yielding small quantities of metal. However, titanium has an exceptionally good strength to weight ratio and is resistant to corrosion. This makes it an ideal engineering material for aircraft construction.
- A cheaper, blast-furnace method cannot be used because the formation of carbides diminishes the desirable properties of the metal.

Make sure you are familiar with the following key terms:

- *contact process*
- *heat exchange*
- *catalysts*
- *Haber process*
- *compromise conditions*
- *recycling*
- *Ostwald process*
- *uses of H_2SO_4, NH_3 and HNO_3*
- *fertilisers*
- *urea*
- *NPK content*
- *blast furnace*
- *chemical reduction*
- *purification of bauxite*
- *electrolysis of Al_2O_3*
- *reduction of $TiCl_4$*
- *properties and uses of iron, steels, aluminium, titanium*

30 mins
Time Yourself

Have you improved?

1 a) Carbon and its compounds are used as both fuel and reducing agent during the extraction of iron. Explain this statement, giving equations where relevant.

C forms CO_2 and CO

b) Explain in terms of costs and its properties why iron is widely used in the construction industry.

Availability

c) What is the main disadvantage of iron and how can it be overcome?

Corrosion

d) Titanium is extracted by reduction of its chloride. Give the equation and conditions for this reduction and explain why titanium is not produced in a similar manner to iron.

$TiCl_4$ + Na

2 a) Outline, giving equations, how pure Al_2O_3 is produced from bauxite.

+NaOH then + CO_2

b) Give the conditions and electrode reactions for the electrolysis of Al_2O_3.

Al^{3+} + O^{2-} ions

c) Electrolysis is expensive in terms of energy. Explain why a cheaper chemical reductant cannot be used to produce aluminium and how these costs influence the site of production.

Cost of electricity

3 a) State the operating temperature and pressure used in the manufacture of NH_3 and explain why they are considered to be compromise conditions.

Balance yield and rate

b) Write an equation for the formation of NH_3 and explain why even though a relatively low conversion is achieved, the overall yield is economic.

Recycling

c) State the role of a catalyst and give the catalyst for the production of NH_3. Why is it essential to purify the feedstock materials?

Rate of reaction

d) Ammonia is converted into nitric acid. Describe the first stage of this process and state two uses of nitric acid.

NO formed

Basic Organic Chemistry and Alkanes

How much do you know?

1 What two elements are always present in hydrocarbons?

2 What happens to boiling points as hydrocarbon chains become longer?

3 What feature of an organic molecule controls its chemical reactivity?

4 What are substances with the same molecular formula but different structure called?

5 Name the molecule with the structural formula CH_3CH_2OH.

6 Draw the structural formula of butane.

7 From what naturally occurring substance are alkanes obtained?

8 What two products do alkanes form when they are combusted in air?

9 Complete the equation $CH_4 + Cl_2 \rightarrow$?

10 What is the name of the mechanism for the reaction above?

Answers

1 carbon and hydrogen **2** increases **3** functional group **4** structural isomers **5** ethanol **6** $CH_3CH_2CH_2CH_3$ **7** crude oil **8** CO_2 and H_2O **9** $\rightarrow CH_3Cl + HCl$ **10** free radical substitution

If you got them all right, skip to page 65

Basic Organic Chemistry and Alkanes

Spend no more than **30 mins** on this topic

Learn the key facts

1 Organic carbon-compounds are based on hydrocarbon chains and rings containing covalently bonded carbon and hydrogen atoms. Straight chain compounds exist as members of homologous series which have :

- the same carbon skeleton;
- chain length increasing by CH_2;
- the same functional groups;
- a gradual trend in physical properties with chain length;
- the same or similar chemical properties.

CH_4
methane

CH_3 $\underline{CH_3}$
ethane

CH_3 $\underline{CH_2}$ CH_3
propane

2 Physical properties are controlled by chain length. As the chain increases, inter-molecular van der Waals forces increase leading to higher boiling points. Longer, non-polar hydrocarbon chains reduce solubility in polar solvents (e.g. H_2O) but increase solubility in non-polar solvents (e.g. hexane).

3 Hydrocarbon chains contain strong C–C and C–H sigma bonds leading to kinetic stability and low chemical reactivity. Chemical properties are governed by functional groups – usually an atom or group of atoms other than carbon and hydrogen. These groups contain areas of charge density such as multiple or polar bonds, which attract and react with other chemical species. Organic compounds are classed by their functional groups.

4 Organic compounds can exist as structural isomers – molecules with the same molecular formula (the number of each type of atom in one molecule) but different structural formula (how the atoms are joined together). E.g. propanone and propanal both have the molecular formula (C_3H_6O) and are structural isomers as shown.

```
    H   O   H                 H     H     O
    |   ||  |                 |     |     ||
H—C — C — C —H        H — C — C — C — H
    |       |                 |     |
    H       H                 H     H
```

propanone C_3H_6O propanal C_3H_6O

1 2 3 4 DAY 5 6 7

5 Organic molecules are named according to systematic rules:

- The longest straight chain gives the root of the name.
- The functional group gives the suffix of the name.
- Any side chains appear as a prefix.
- Side chains and functional group chain positions are given by numbers.

Chain Length	*Name*	*Functional Group*	*Prefix*	*Suffix*
1	Methan-	Alkane, C–H	alkyl	-ane
2	Ethan-	Alkene, C=C		-ene
3	Propan-	Haloalkane, C–X	halo-	
4	Butan-	Alcohol, OH	hydroxy-	-ol
5	Pentan-	Acid, COOH		-oic acid
6	Hexan-	Ketone, COC	oxo	-one
		Aldehyde, RCHO		-al

```
  H   H   H
  |   |   |
H—C — C — C — Cl
  |   |   |
  H   H   H
1 – chloropropane

    H   Cl  H
    |   |   |
H — C — C — C — H
    |   |   |
    H   H   H
2 – chloropropane

    H    H   H   H
    |    |   |   |
H — C —  C — C — C—OH
    |    |   |   |
   CH3   H   H   H
2 – methyl butan–1–ol

    H         O
    |        //
H — C  —  C
    |        \
    H         OH
ethanoic acid
```

6 Alkanes are saturated hydrocarbons of general formula C_nH_{2n+2}. Because alkanes have no functional groups, they are relatively chemically inert. They show structural isomerism in the form of chain isomerism, e.g. pentane, C_5H_{10} has three structural isomers.

```
                                            CH3
                                            |
CH3CH2CH2CH3      CH3CH2CHCH3      CH3 — C — CH3
                        |                   |
                        CH3                 CH3

pentane           2–methylbutane   2,2 dimethyl propane
```

7 Alkanes are obtained as simple mixtures, according to chain length, by fractional distillation of crude oil and have numerous commercial uses.

Fraction	Carbon content	Uses
Light Petroleum Gas	C_1–C_4	gas fuels, organic chemicals
Gasoline	C_5–C_{10}	petrol, organic chemicals
Kerosene	C_{10}–C_{16}	jet fuel
Diesel	C_{13}–C_{25}	diesel fuel, heating fuel
Residue	$>C_{25}$	bitumen, lubrication, waxes

Long chain hydrocarbons are converted into valuable shorter chains by catalytic cracking using silica or alumina at 400–500 °C (free radical reaction).

$C_{16}H_{34} \rightarrow C_8H_{18} + 4C_2H_4$ (octane and ethene)

8 Alkanes are used as fuels because they combust exothermically in excess air to form CO_2 and H_2O, e.g. petrol (octane), butane and methane gases. Liquid fuels have the advantage of being easier to store and transport.

$C_8H_{18} + 12\frac{1}{2}O_2 \rightarrow 8CO_2 + 9H_2O \quad \Delta H = -5510\,kJ\,mol^{-1}$

Pollution from combustion engines is a major environmental problem.

	Effect	Formation	Equations
CO	toxic	partial combustion	$C_8H_{18} + 8^1/_2O_2 \rightarrow 8CO + 9H_2O$
NO	photo-chemical smog/acid rain	combustion of N_2 in air	$N_2 + O_2 \rightarrow 2NO$ Then $2NO + O_2 \rightarrow 2NO_2$
C_8H_{18}	carcinogenic	unburned petrol	

Many emissions can be removed by platinum/rhodium catalytic converters.

$2CO + 2NO \rightarrow 2CO_2 + N_2$ and $C_8H_{18} + 25NO \rightarrow 8CO_2 + 9H_2O + 12\frac{1}{2}N_2$

1
2
3
4
DAY
5
6
7

9 Alkanes react with Cl_2 and Br_2 under ultraviolet light to form haloalkanes via free radical substitution. Free radicals are reactive species with an unpaired electron. Bi-products such as ethane or dichloromethane, CH_2Cl_2 (further substitution), often form so yields of pure CH_3Cl are low. Reaction products are separated by fractional distillation.

$$CH_4 + Cl_2 \rightarrow CH_3Cl + HCl$$

10 The mechanism for this reaction is as follows:

Initiation: $Cl_2 \rightarrow 2Cl\cdot$ (UV light breaks Cl–Cl bond to form radicals)

Propagation: $CH_4 + Cl\cdot \rightarrow CH_3\cdot + HCl$, $CH_3\cdot + Cl_2 \rightarrow CH_3Cl + Cl\cdot$

Termination: $CH_3\cdot + Cl\cdot \rightarrow CH_3Cl$ and $Cl\cdot + Cl\cdot \rightarrow Cl_2$

Bi-products can be formed as follows:

Propagation: $CH_3Cl + Cl\cdot \rightarrow CH_2Cl\cdot + HCl$,

Further · substitution: $CH_2Cl\cdot + Cl_2 \rightarrow \underline{CH_2Cl_2} + Cl$

Termination: $CH_3\cdot + CH_3\cdot \rightarrow \underline{CH_3CH_3}$

Make sure you understand the following terms:

- *hydrocarbon*
- *homologous series*
- *trends in physical properties*
- *functional groups*
- *structural isomerism*
- *systematic naming rules*
- *alkane*
- *fractional distillation*
- *cracking*
- *combustion*
- *pollution*
- *catalytic converter*
- *chlorination*
- *free radical substitution mechanism*
- *initiation, propagation, termination*
- *bi-products of chlorination*

Have you improved?

1 Many organic compounds exist as homologous series such as straight chain alcohols which have the general formula $C_nH_{2n+1}OH$.

a) State three similarities of the members of this homologous series.

Members only differ by CH_2

b) Explain why ethanol, C_2H_5OH, is soluble in water but larger alcohols are less so.

Hydrocarbon non-polar

c) What other physical property changes down the series? Explain why it does so.

Increasing intermolecular forces

d) Butan-1-ol has the structural formula $CH_3CH_2CH_2CH_2OH$. Give the name and formula of two structural isomers of this molecule.

Move the OH or CH_3

2 Alkanes are saturated hydrocarbons which are relatively unreactive.

a) Explain briefly why alkanes are relatively unreactive but *are* widely used as fuels when combusted.

C and H from strong bonds

b) Write an equation for the *complete* combustion of heptane, C_7H_{16} and explain why incomplete combustion can lead to environmental pollution.

CO_2 and CO formed

c) Nitrogen monoxide is emitted from car exhausts and also leads to environmental pollution. State why it is a pollutant and describe, giving an equation, how its emissions can be reduced.

NO converted to N_2

3 Methane reacts with chlorine gas under ultraviolet light via a free radical substitution reaction. The overall equation is $CH_4 + Cl_2 \rightarrow CH_3Cl + HCl$ but the reaction proceeds through several steps.

a) What is the name of the first step of this reaction and why is ultraviolet light required?

Cl· radicals required

b) Give the propagation stages of the reaction to show how CH_3Cl is formed.

$CH_4 + Cl\cdot \rightarrow$? etc.

c) When the final reaction mixture was analysed it was found to contain significant quantities of a molecule with an Mr of 85. Suggest the identity of this species, how it was formed and how it might be separated from the mixture.

Further substitution occurs

Alkenes

How much do you know?

1 What is the functional group of alkenes?

2 What type of isomerism occurs because of the functional group in question 1?

3 Name the product formed when ethene reacts with HBr.

4 What is observed when alkenes react with a solution of bromine?

5 What type mechanism does the reaction in question 3 undergo?

6 What is the major product when propene reacts with HBr?

7 What kind of reaction do alkenes undergo with a solution of alkaline potassium manganate?

8 Name the molecule $(CH_2)_2O$.

9 What type of reaction occurs during the production of margarine?

10 What type of polymers do alkenes form?

Answers

1 carbon–carbon double bond **2** geometric **3** bromoethane **4** red-brown decolourises **5** electrophilic addition **6** 2-bromopropane **7** oxidation **8** epoxyethane **9** hydrogenation **10** addition

If you got them all right, skip to page 71

Alkenes

Spend no more than **30 mins** on this topic

Learn the key facts

1 Alkenes are hydrocarbon chains which contain carbon–carbon double bonds, C=C. These consist of a sigma bond and a pi-bond, the latter formed by the overlap of p-orbitals producing regions of electron density above *and* below the sigma bond. A double bond is a relatively high region of negative charge (attracting electrophiles) and the pi-bond is relatively weak, easily breaking during reactions. Thus alkenes are more reactive than alkanes.

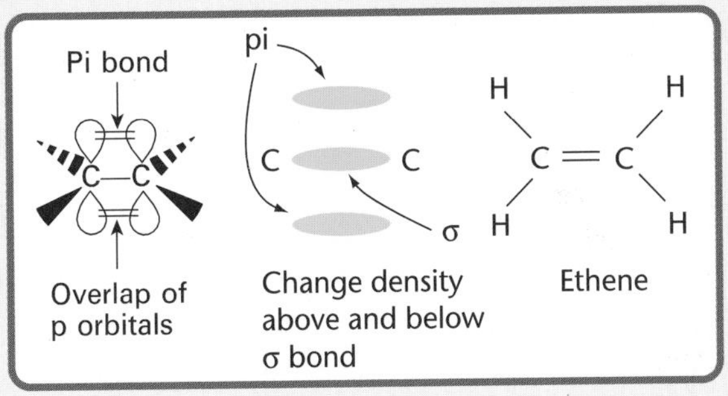

Pi bond

Overlap of p orbitals

pi

σ

Change density above and below σ bond

Ethene

2 The double bond can occupy different positions on the carbon skeleton, leading to structural isomerism. Double bonds cannot rotate (unlike single bonds) without breaking the pi-component. This can lead to geometric isomerism (a form of stereo-isomerism) where groups are fixed in position either on the same side of the double bond (cis) or on opposite sides (trans).

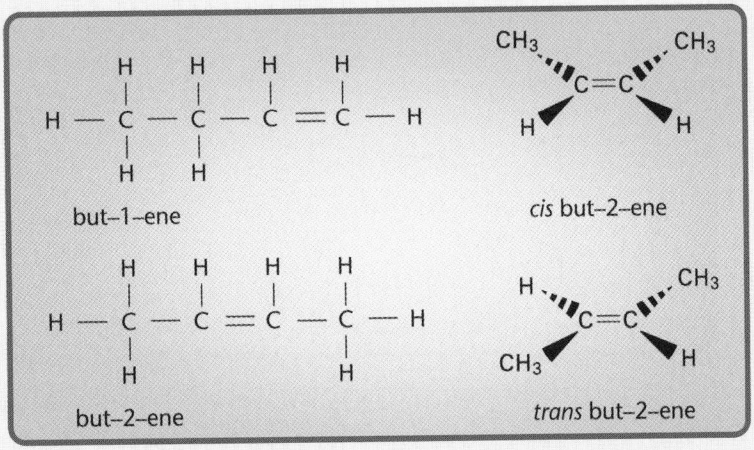

but–1–ene

cis but–2–ene

but–2–ene

trans but–2–ene

DAY 6

3 Alkenes undergo electrophilic addition across the double bond to produce a single product. The negative charge density of the double bond attracts electrophiles – species capable of accepting a pair of electrons. Common electrophiles are HBr (gas or concentrated solution at room temperature) and H_2SO_4 (concentrated at room temperature) which have polar bonds giving rise to an electrophilic, δ^+ atom. The pi bond breaks and the whole of the 'attacking' species is added, saturating the alkene.

$CH_2 = CH_2 + HBr \rightarrow CH_3CH_2Br$ (ethene to bromoethane)

$CH_3CH = CH_2 + H{-}OSO_3H \rightarrow CH_3CH(OSO_3H)CH_3$

4 Bromine, dissolved in carbon tetrachloride (CCl_4) or water, also adds across the double bond (a dipole is induced into the Br_2 molecule as it approaches the double bond) to produce di-bromoalkanes. The red-brown colour of the bromine is decolourised and so this reaction can be used as a test for the presence of carbon–carbon double bonds. If bromine is added quantitatively the number of double bonds can also found.

$C_2H_5CH{=}CHCH_3 + Br_2 \rightarrow C_2H_5CHBrCHBrCH_3$ (1,2-dibromopentane)

5 The electrophilic addition mechanism is shown below. Note that the pi-bond breaks heterolytically because both electrons go to the same species.

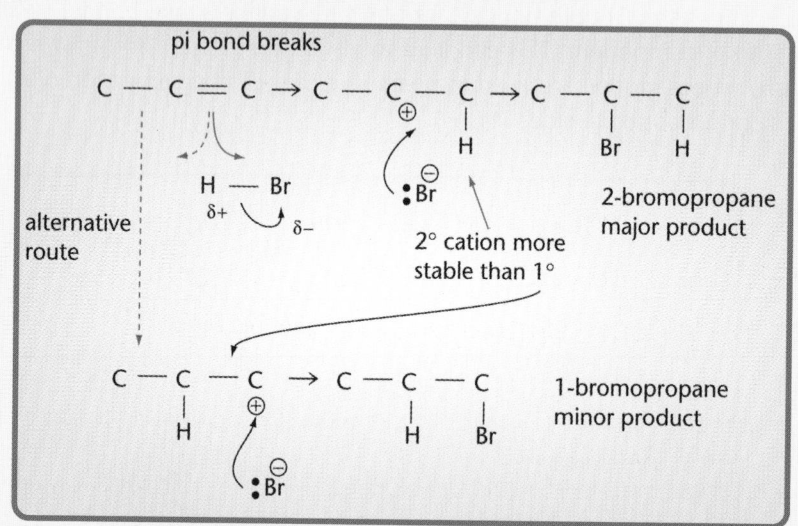

6 Markovnikov's rule states that in the addition of hydrogen halides to unsymmetrical alkenes the hydrogen adds to the least substituted carbon. This allows the reaction to proceed via the most substituted carbocation intermediate which is stabilised by the greatest number of positive inductive effects from the bonded alkyl groups. Thus tertiary is more stable than secondary, which is more stable than primary. Therefore in the addition of HBr to propene, *2*-bromopropane is the major product and *1*-bromopropane the minor product (as shown).

7 Alkenes are oxidised by cold, aqueous $KMnO_4$ and NaOH to give diols.

$CH_2{=}CH_2 + [O] + H_2O \rightarrow CH_2(OH)CH_2OH$ (ethene to ethane-1,2-diol)

8 Ethene can be oxidised by oxygen (air) using a silver catalyst to produce epoxyethane, containing a three member C-O-C ring.

$2CH_2{=}CH_2 + O_2 \rightarrow 2(CH_2)_2O$ (epoxyethane)

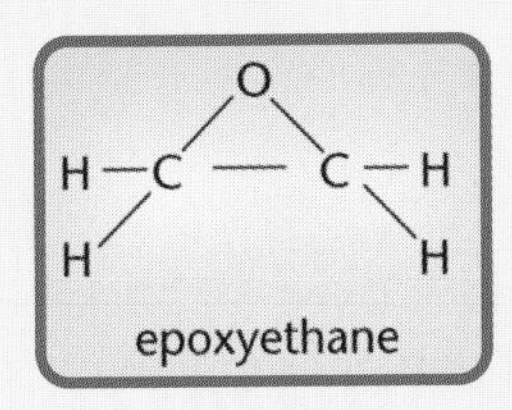

epoxyethane

The epoxide ring contains small bond angles (<109°) causing bond strain and hence reactivity. The ring is readily opened by heating with H_2O (hydrolysis), to form ethane-1,2-diol, and with alcohols to form a range of useful materials.

$(CH_2)_2O + H_2O \rightarrow HOCH_2CH_2OH$ – (antifreeze and polyesters)

$(CH_2)_2O$ + x's $CH_3OH \rightarrow CH_3OCH_2CH_2OH$ (paints and solvents)

$n(CH_2)_2O + CH_3OH \rightarrow CH_3O(CH_2CH_2O)_nH$ (plasticisers)

9 Catalytic hydrogenation using H_2 gas and a Ni catalyst at 150 °C is used to convert unsaturated vegetable oils into saturated, spreadable fats such as margarine. Acid-catalysed hydration using steam (H_2O) and a phosphoric acid catalyst at 300 °C and 70 atm is used to produce ethanol from ethene. This process is faster and gives a better, purer yield of ethanol compared to traditional fermentation processes.

$RCH{=}CHR + H_2 \rightarrow RCH_2CH_2R$ (margarine)

$CH_2{=}CH_2 + H_2O \rightarrow CH_3CH_2OH$ (ethanol)

10 Alkenes undergo free radical (homolytic) addition polymerisation to form unsaturated polymer chains consisting of many thousands of repeating units – the smallest repeating structure. Examples of addition polymers include polyethene, polypropene, polychloroethene (PVC) and polytetrafluoroethene (PTFE or Teflon).

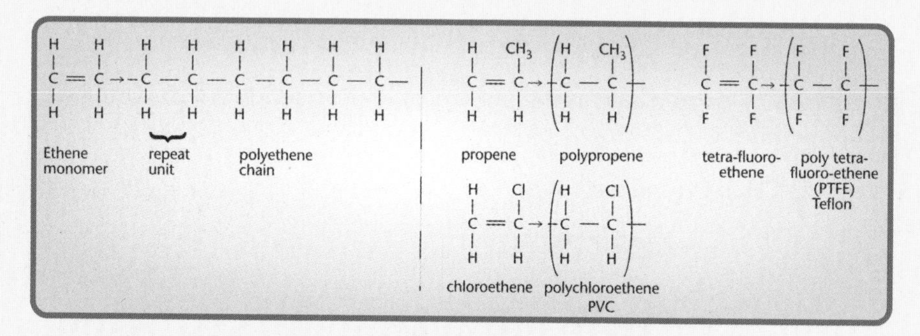

- Polyethene: soft, light and flexible. Used for packaging and plastic bags.
- Polypropene: tougher and with a higher melting point than polyethene. Used for ropes and packaging.
- Polychloroethene: rigid and chemically resistant. Used as building materials such as for guttering and electrical insulation.
- Polytetrafluoroethene: has a smooth, 'slippery' surface. Used for non-stick surfaces for saucepans and as washers for plumbing.

Make sure you understand the following terms:

- *alkene*
- *double bond*
- *p-orbitals*
- *alkene reactivity*
- *positional isomerism*
- *geometric isomerism*
- *electrophilic addition with HBr and H_2SO_4*
- *electrophile*
- *chemical test with Br_2*
- *(heterolytic) electrophilic addition mechanism*
- *Markovnikov addition*
- *carbo-cation intermediate*
- *oxidation*
- *epoxyethane*
- *hydrogenation*
- *hydration*
- *self-addition polymerisation*
- *properties of polymers*

Have you improved?

1 a) Describe, using suitable diagrams, what is meant by a pi-bond which is present in carbon–carbon double bonds.

P-orbitals

b) Explain why alkenes are more reactive than alkanes.

Pi-bond weak

c) Draw the geometric isomers of but-2-ene and explain how this type of isomerism occurs.

Groups opposite or adjacent

2 a) Write an equation for the reaction between but-2-ene and hydrogen bromide, name the product and state the reaction conditions necessary.

HBr adds to propene

b) Draw the mechanism to show the formation of the *major* product between propene and hydrogen bromide.

Pi-bond breaks to bond with H-atom

c) Explain why the product you have stated in b) is formed preferentially.

Carbocation stability

3 a) Describe how you would carry out a chemical test to show the presence of a carbon–carbon double bond. What would you observe?

Decolourisation

b) Write an equation to show the formation of epoxyethane from ethene and then the conversion of epoxyethane into ethane-1,2-diol. What is this final product used for?

Epoxyethane is $(CH_2)_2O$

c) Draw the structure of the polytetrafluoroethene monomer and a section of the addition polymer it forms showing at least three repeating units. What type of reaction does addition polymerisation proceed by? Suggest a use for this polymer.

Pi-bond breaks and monomers add together

Haloalkanes

How much do you know?

1 What structural feature of haloalkanes leads to their reactivity?

2 What kind of reactions do haloalkanes undergo with nucleophiles?

3 Which undergoes substitution more quickly – chloroethane or iodoethane?

4 Write an equation for the reaction between 1-bromopropane and NaOH.

5 What reagents would you use to test for the presence of a chloroalkane?

6 Identify the organic product when bromoethane reacts with KCN?

7 What reagents and conditions are required to convert CH_3Br to CH_3NH_2?

8 State the role of the OH^- ion when haloalkanes undergo elimination with KOH in ethanol.

9 What type of haloalkanes are used as refrigerants?

Answers

1 polar C–X bond **2** substitution **3** iodoethane **4** $C_3H_7Br + NaOH \rightarrow C_3H_7OH + NaBr$ **5** HNO_3, $AgNO_3$ (aq) **6** propanitrile **7** NH_3 in ethanol plus heat **8** base **9** chlorofluorocarbons (CFCs)

If you got them all right, skip to page 76

Spend no more than 30 mins on this topic

Learn the key facts

1 Haloalkanes are saturated compounds that contain polar, covalent carbon–halogen bonds, C–X. The δ+ carbon atom is susceptible to attack by nucleophiles: electron-pair donors such as OH^-, CN^- and NH_3.

2 Primary haloalkanes react with nucleophiles via a nucleophilic substitution mechanism (shown for the OH^- ion). The nucleophile approaches the δ+ carbon opposite to the δ– halogen, and forms a dative bond. At the same time the C–X bond is broken, releasing the halide ion.

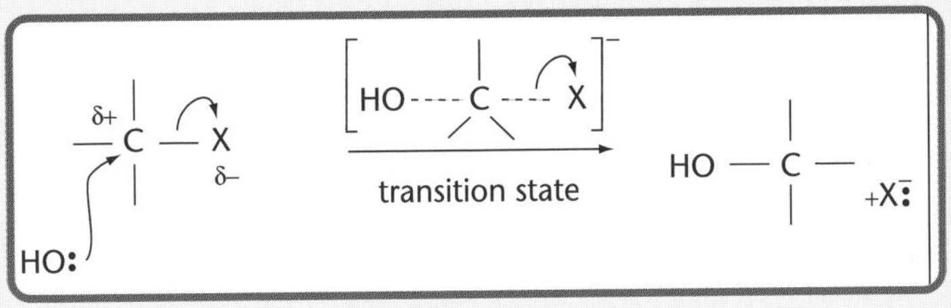

δ+ C — X δ–
HO:
[HO---- C ---- X]⁻
transition state
HO — C — +X:⁻

3 As the halogen atom becomes larger the C–X bond becomes weaker and is more easily broken, lowering the activation energy for substitution. Thus iodoalkanes are substituted more quickly than chloroalkanes.

4 Haloalkanes react with warm, aqueous NaOH to produce alcohols via nucleophilic substitution. The reaction is known as hydrolysis and the OH^- ions behave as nucleophiles.

$CH_3CH_2Cl + NaOH \rightarrow CH_3CH_2OH + NaCl$ (chloroethane to ethanol)

5 Haloalkanes are tested for by identifying the halide ion released by hydrolysis. The solution is neutralised with nitric acid and aqueous silver nitrate, $AgNO_3$, added to give coloured, silver-halide precipitates, AgX. These are then distinguished by their solubility in aqueous ammonia.

Haloalkane	***ppt***	***+ dil. NH_3***	***+ conc NH_3***
RCl	AgCl white	soluble	soluble
RBr	AgBr cream	insoluble	soluble
RI	AgI yellow	insoluble	insoluble

6 When heated with potassium cyanide in aqueous ethanol, haloalkanes undergo nucleophilic substitution to form nitriles, RCN. The CN^- ion acts as the nucleophile. This reaction is important because the carbon skeleton has been extended by 1. The nitrile group can then be hydrolysed by heating under reflux with dilute aqueous H_2SO_4 to produce a carboxylic acid.

$C_2H_5Br + KCN \rightarrow C_2H_5CN + KBr$ (bromoethane to propanitrile)

$C_2H_5CN + 2H_2O \rightarrow C_2H_5CO_2H + NH_3$ (propanitrile to propanoic acid)

7 Heating with alcoholic NH_3 converts haloalkanes into primary amines, RNH_2, via nucleophilic substitution.

$CH_3Br + NH_3 \rightarrow CH_3NH_2 + HBr$ (bromomethane to methylamine)

Because amines also posses a lone pair on the N atom they are also nucleophiles and in excess NH_3 further substitution can occur, leading to secondary and tertiary amines, R_2NH and R_3N and quaternary ammonium salts, $R_4N^+X^-$.
Long chain quaternary salts are used as cationic detergents.

$CH_3NH_2 + CH_3Br \rightarrow (CH_3)_2NH + HBr$ (secondary amine)

$(CH_3)_3N + CH_3Br \rightarrow (CH_3)_4N^+Br^-$ (quaternary ammonium salt)

8 Heating with KOH in alcohol converts haloalkanes to alkenes via an elimination reaction. The OH^- ion behaves as a base and NOT a nucleophile. An H can be lost from either side of the carbon bearing the halogen, yielding two isomeric alkenes. (Note that regardless of reaction conditions, elimination and substitution usually both occur as competing reactions.)

$C_2H_5Cl + KOH \rightarrow C_2H_4 + H_2O + KBr$ (chloroethane to ethene)

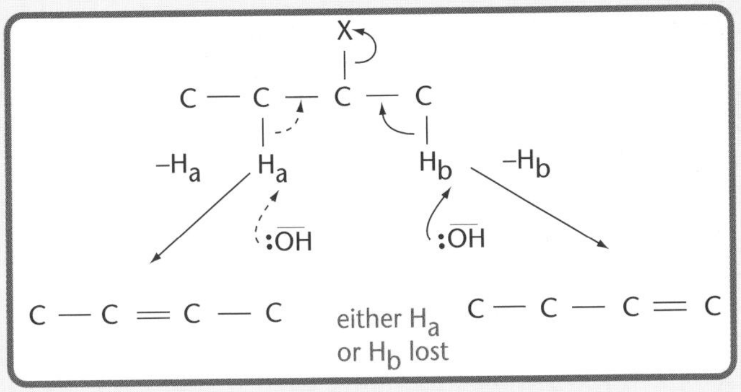

X
C — C — C — C
$-H_a$ H_a H_b $-H_b$
$:\overline{OH}$ $:\overline{OH}$
C — C = C — C
either H_a or H_b lost
C — C — C = C

9 Chlorofluorocarbons (CFCs) are used as refrigerants, cleaning agents and propellants. However, their use is being reduced due to their role in the depletion of the ozone layer.

Chloroalkanes are widely used as pesticides and solvents. However, they are often toxic and non-biodegradable, which leads to health risks if they are retained in the food chain or water system.

Make sure you understand the following key points:

- *haloalkanes*
- *polar C–X bond*
- *nucleophile*
- *nucleophilic substitution mechanism*
- *rate of substitution*
- *hydrolysis*
- *test for haloalkanes*
- *nitriles*
- *amines and further substitution*
- *elimination*
- *uses of haloalkanes*

Have you improved?

1 a) Give suitable reagents and conditions for the conversion of 1-bromobutane into butan-1-ol and write and equation for this reaction.

b) Draw the mechanism for the reaction above and state the role of the OH^-ion.

c) Explain why iodobutane is found to react more quickly than bromobutane under the conditions you have described.

2 a) Write an equation for the reaction that occurs when chloroethane is heated with ammonia in alcohol to produce ethylamine.

b) Explain why other organic products are also produced.

c) Give the formula of the likely final product formed if a large excess of haloalkane is used. What type of compound is this and what commercial use does it have?

3 a) Draw the mechanism for the reaction when 1-bromobutane is heated in alcoholic KOH. Name the mechanism and state the role of the OH^- ion.

b) Explain why two structurally isomeric products are formed when 2-bromobutane is reacted in the same way. Give the structures of the products.

c) Give a commercial use of chloroalkanes and explain why they pose an environmental hazard.

Need OH^- ions

OH^- uses its lone pair to attack δ^+ C-atom

Bond strength

Cl replaced by NH_2

Further substitution

N forms 4 bonds

OH- removes H atom

H can be removed from two sites

Non-biodegradable

Alcohols, Carbonyls and Acids

How much do you know?

1 Write the structural formula of propan-2-ol.

2 What is observed when ethanol is oxidised using $KMnO_4(aq)$ and dil. $H_2SO_4(aq)$?

3 Complete the equation $CH_3CH_2OH + 2[O] \rightarrow$?

4 Name the organic product formed when butan-2-ol is oxidised.

5 What is observed when propanal is treated with Tollen's reagent?

6 Complete the equation $CH_3CHO + 2[H] \rightarrow$?

7 What type of reaction occurs when alcohols are treated with hot, concentrated H_2SO_4?

8 Suggest reagents and conditions to covert methanol to bromomethane.

9 What type of organic compound is formed when an alcohol is boiled with a carboxylic acid and concentrated H_2SO_4?

10 By what process are alcohols produced from glucose?

Answers

1 $CH_3CH(OH)CH_3$ **2** purple to colourless **3** $\rightarrow CH_3COOH + H_2O$ **4** butanone **5** a silver metal ppt or mirror **6** $\rightarrow CH_3CH_2OH$ **7** elimination or dehydration **8** NaBr & conc. H_2SO_4, heating **9** ester **10** fermentation.

If you got them all right, skip to page 81

Spend no more than 30 mins on this topic

Alcohols, Carbonyls and Acids

Learn the key facts

1 Alcohols are saturated compounds containing polar OH groups. They have high boiling points and are soluble in water due to hydrogen bonding between OH groups. They are classified as primary (1°), secondary (2°) and tertiary (3°) based on the number of carbons bonded to the C-atom bearing the OH group.

$$CH_3-CH_2-OH \quad \text{1° ethanol}$$

$$CH_3-CH(OH)-CH_3 \quad \text{2° propan–2–ol}$$

$$CH_3-C(OH)(CH_3)-CH_3 \quad \text{3° 2-methyl propan–2–ol}$$

2 Alcohols are oxidised by heating with aqueous $KMnO_4$ (turns purple to colourless) or $K_2Cr_2O_7$ (turns orange to green) with dilute H_2SO_4.

3 1° alcohols oxidise to form carboxylic acids.

$C_2H_5OH + 2[O] \rightarrow CH_3COOH + H_2O$ (ethanol to ethanoic acid)

The intermediate aldehyde, RCHO, is isolated by adding the oxidant dropwise and distilling off the aldehyde.

$C_2H_5OH + [O] \rightarrow CH_3CHO + H_2O$ (ethanol to ethanal)

$$-CH_2-OH \rightarrow -C(=O)OH$$

$$-CH_2-OH \rightarrow -C(=O)-H$$

4 2° alcohols oxidise to form ketones.

$CH_3CH(OH)CH_3 + [O] \rightarrow CH_3COCH_3 + H_2O$ (propan-2-ol to propanone)

3° alcohols do not oxidise.

$$-CH(OH)-C \rightarrow -C(=O)-C$$

DAY 7

5 Aldehydes are distinguished from ketones (both are carbonyls containing the C=O group) by Fehling's reagent (Cu^{2+}) or Tollen's reagent ($AgNO_3/NH_3$). Aldehydes are oxidised to the carboxylic acid and react positively.

Tollens: $Ag^+ \rightarrow Ag^0$, a silver ppt or 'mirror' observed.

Fehling's: $Cu^{2+} \rightarrow Cu_2O$ (Cu(I)), the blue solution turns to a brick-red ppt.

6 Aldehydes and carboxylic acids are reduced to 1° alcohols by lithium aluminium hydride, $LiAlH_4$, in dry ether. Aldehydes can also be reduced by sodium borohydride, $NaBH_4$, in ethanol. Reduction is the addition of H or the removal of O. Ketones reduce to 2° alcohols.

$C_3H_7COOH + 4[H] \rightarrow C_3H_7CH_2OH + H_2O$ (butanoic acid to butan-1-ol)

$C_3H_7CHO + 2[H] \rightarrow C_3H_7CH_2OH$ (butanal to butan-1-ol)

$CH_3COCH_3 + 2[H] \rightarrow CH_3CH(OH)CH_3$ (propanone to propan-2-ol)

7 Alcohols undergo elimination with conc. H_2SO_4 at 180 °C to form alkenes. The strong acid protonates the OH group, followed by the elimination of water (dehydration) and then H^+. Depending which H^+ is lost, two isomeric alkenes can form. Al_2O_3 at 350 °C can also be used.

H⁺ O — H
C — C — C — C → C — C(O⁺H₂) — C — C → C — C(Ha) — C⁺ — C(Hb)
–Hb → C — C — C = C
–Ha → C — C = C — C

1
2
3
4
5
6
DAY 7

8 The OH group can be substituted by a halogen atom as follows.

Chlorination: dry PCl_5 at room temperature.

$CH_3OH + PCl_5 \rightarrow CH_3Cl + POCl_3 + HCl$ (methanol to chloromethane)

This is a test for the OH group (including carboxylic acids). The misty fumes of HCl gas produced turn damp blue litmus paper red.

Bromination: concentrated H_2SO_4 + NaBr to form HBr *in situ*, heating.

$C_2H_5OH + HBr \rightarrow C_2H_5Br + H_2O$ (ethanol to bromoethane).

Iodination: P + I_2 to form PI_3 *in situ*, heating.

$3C_3H_7OH + PI_3 \rightarrow 3C_3H_7I + H_3PO_3$ (propan-1-ol to 1-iodopropane).

9 Alcohols react with sodium metal at RT to form alkoxides. They form esters when boiled with carboxylic acids and a concentrated H_2SO_4 catalyst.

$C_2H_5OH + Na \rightarrow C_2H_5O^-Na^+ + \frac{1}{2}H_2$ (gas) (ethanol to sodium ethoxide).

10 $CH_3COOH + CH_3OH \rightarrow CH_3COOCH_3 + H_2O$ (methanol to methylethanoate).

Ethanol can be cheaply produced by fermentation of natural, renewable glucose (e.g. sugar beet or cane) and is often used as a liquid fuel instead of petrol in combustion engines. This method of production gives a relatively low, impure yield but exploits a resource that is renewable and cheap.

$C_6H_{12}O_6 \rightarrow 2C_2H_5OH + 2CO_2$ (35 °C, yeast catalyst)

$C_2H_5OH + 3O_2 \rightarrow 2CO_2 + 3H_2O$ (combustion)

Make sure you understand the following key points:

- *OH group*
- *H-bonding*
- *oxidation of 1° 2° and 3° alcohols*
- *Fehling's solution & Tollen's reagent*
- *reduction of aldehydes, ketones and carboxylic acids*
- *$LiAlH_4$*
- *dehydration*
- *halogenation*
- *test for OH*
- *manufacture of ethanol*

Have you improved?

1 a) Draw the displayed structures of a primary, secondary and tertiary alcohol which all have the molecular formula $C_4H_{10}O$. Name these alcohols. *Change the carbon skeleton*

b) How you would test for the presence of the OH group? Describe what you would observe. *PCl_5*

c) State what would occur when each alcohol is heated separately with an excess of $K_2Cr_2O_7$ (aq) and dilute H_2SO_4. Name any organic products formed. *Which alcohols oxidise?*

d) Write a balanced equation for the reaction of the primary alcohol in c). *Add O and remove H*

2 a) Give the structural formula for butanal and butanone. *Both have C=O*

b) Suggest a suitable reagent for the reduction of these substances and write a balanced equation for the reduction of butanone. *Addition of H*

c) Describe and explain how Fehling's solution can be used to distinguish between butanal and butanone. *Aldehydes oxidised*

3 a) Butan-2-ol undergoes dehydration when heated with concentrated H_2SO_4. Write an equation for this reaction and explain what type of reaction it is. *H_2O removed*

b) Name the two isomeric, organic products which could be formed when butan-2-ol is dehydrated and explain why they are formed. *H^+ removed from two sites*

c) State the function of the H_2SO_4 in the first stage of this reaction. What other reagents and conditions could be used to achieve dehydration? *Protonation*

Exam Practice Questions

1 A 0.15 g sample of impure sodium chloride was dissolved in 25 cm^3 of distilled water. This solution was titrated with 0.1 M silver nitrate solution. The titration required 23.40 cm^3 of silver nitrate.
The equation for the reaction occurring is:

$$NaCl_{(aq)} + AgNO_{3(aq)} \rightarrow AgCl_{(s)} + NaNO_{3(aq)}$$

a) Calculate the number of moles of silver nitrate reacting.
b) How many moles of sodium chloride were dissolved?
c) What mass of sodium chloride was dissolved?
d) Calculate the percentage of sodium chloride in the impure sample.
e) Explain why a solid was observed during the reaction.
f) Write an ionic equation for the reaction occurring.

2 This question concerns the following reaction:

$$CH_4 + 2O_2 \rightarrow CO_2 + 2H_2O$$

You may find the following information useful:

Substance	ΔH_f^θ kJ mol^{-1}
CH_4	−75
CO_2	−394
H_2O	−286

a) What is represented by ΔH_f^θ?
b) Define this term.
c) What are the standard conditions?
d) Using the information provided construct an energy cycle and use it to determine the enthalpy change for the reaction.
e) In light of the above answer, explain why methane does not spontaneously combust in air.

1 2 3 4 5 6 DAY 7

3 Consider the following reaction scheme

$$C_4H_8\ A \xrightarrow{HBr} B \xrightarrow{NaOH} C \xrightarrow[+\text{dil } H_2SO_4]{K_2Cr_2O_7} D,\ C_4H_8O$$

C $\xrightarrow{\text{conc } H_2SO_4}$ A + E

E is a structural isomer of A.

a) Give the structural formulae and names for A – E.
b) What type of isomerism do A and E both show and how does it occur?
c) F is a structural isomer of D which does not decolourise a solution of Br_2 in CCl_4 and has a branched chain. Draw the structure of F and describe a chemical test to distinguish it from D.

4 a) Write an equation for the reaction between phosphorous and chlorine to form phosphorous trichloride. What kind of reaction is this?
b) Explain why phosphorous trichloride is found to have a relatively low boiling point and does not conduct electricity.
c) A mixture of two group 2 metal chlorides, W and X, are found to dissolve in water to produce a clear colourless solution. Addition of dilute H_2SO_4 produces a thick white ppt, Y, which is filtered off. The residual solution contains the metal ion Z. When this solution is evaporated, X is reformed and was found to impart a brick red colour to a hot flame. Explain these observations and identify W, X, Y and Z.

Answers on page 93

Have you improved: Answers

Atomic Structure

1 a) $1s^2\ 2s^2\ 2p^6\ 3s^2\ 3p^5$

b) ^{35}Cl protons = 17 neutrons = 18 electrons = 17

^{37}Cl protons = 17 neutrons = 20 electrons = 17

c) Atoms of the same element that have the same number of protons but different numbers of neutrons.

d) $Cl_{(g)} \rightarrow Cl^+_{(g)} + e^-$

e) $Cl^+_{(g)} \rightarrow Cl^{2+}_{(g)} + e^-$

f) The positive chlorine ion has more protons than electrons meaning that the electrons are more strongly attracted to the nucleus.

2 a) ^{10}B = 20%. ^{11}B = 80%

b) mass ÷ charge

c) +1

d) $(10 \times 20) + (11 \times 80) \div 100 = 10.8$

Formulae and Equations

1 a) $C_2H_3O_2$

b) $C_4H_6O_4$

2 a) $Na_2CO_{3\ (aq)} + 2HCl_{\ (aq)} \rightarrow 2NaCl_{(aq)} + H_2O_{(l)} + CO_{2(g)}$

b) $CO_3{}^{2-}{}_{(aq)} + 2H^+{}_{(aq)} \rightarrow H_2O_{(l)} + CO_{2\ (g)}$

3 RMM = 63.57

Moles

1 a) i) 2

ii) 4

iii) 10

b) i) 6.02×10^{23}

ii) 1.204×10^{24}

iii) 9.03×10^{23}

2 The number of moles of hydrochloric acid reacting = 0.048, so the number of moles of calcium carbonate reacting = 0.024. The mass of calcium carbonate reacting = 2.4 g hence the percentage of calcium carbonate in the marble
= 80%.

3 $XOH + HCl$ R $XCl + H_2O$. The number of moles of hydrochloric acid reacting is 0.0325. Hence the number of moles of XOH in 25 cm^3 is 0.0325 and therefore the number of moles in 250 cm^3 is 0.325. This number of moles weighs 18.2 g therefore 1 mole weighs 56 g and the RMM of XOH is 56. Therefore the mass of X is 56 – (16+1) = 39, so X is potassium.

4 $0.2 \times 24\,dm^3 = 4.8\,dm^3$

Structure and Bonding

1 a) 3 b.p. and 1 l.p. which repel each other, therefore pyramidal
b) Based on tetrahedron but l.p. repels more than b.p. therefore approx. 107°.
c) When PCl_3 melts only weak vdW forces are broken so only a small amount of energy required. Particles increase in kinetic energy and vibrate quickly until force of attraction are overcome and particles are free to flow.

2 a)

```
   ••
  •S•
 •×   •×
H       H
```

b) Boiling point decreases from H_2Te to H_2S as central atom becomes smaller so van der Waal's forces decrease. But H_2O contains strong hydrogen bonds which require more energy to break, increasing boiling point.
c) H_2O molecules in ice are further apart in the solid to allow each molecule to form 2 hydrogen bonds as well as 2 covalent bonds, leading to an open, regular tetrahedral-structure.

3 a) Cl atoms are smaller so attraction for bond electrons is greater therefore more energy required to break this bond.
b) Energy required to break down ionic lattice compensated for by the energy released by hydration of ions. Ions are mobile in solution and carry electrical current.
c) Li^+ cation is small and has a high surface charge density, therefore it polarises the large I^- anion, distorting electron density towards the cation. Therefore some electron density is shared.

Have you improved: Answers

The Periodic Table

1 a) Decreases.
b) A smaller atom means that the electrons are closer to the nucleus and are therefore more strongly attracted to the nucleus. This means that more energy is required to remove them.
c) X = Al, Y = Si, Z = P

2 a) Metallic structure which means that there are free/mobile electrons.
b) Very weak intermolecular bonding (van der Waals).
c) Covalent structure therefore there are no free electrons.

Introduction to Oxidation and Reduction

1 a) +6
b) +5
c) +2

2 a) i) C
ii) Zn
iii) Cl^-
b) i) PbO
ii) Fe^{2+}

3 a) $MnO_4^- + 4H^+ + 3e^- \rightarrow MnO_2$
b) $MnO_4^- + 8H^+ + 5e^- \rightarrow Mn^{2+} + 4H_2O$
c) $C_2O_4^{2-} \rightarrow 2CO_2 + 2e^-$
d) $2S_2O_3^{2-} \rightarrow S_4O_6^{2-} + 2e^-$

4 $Cr_2O_7^{2-} + 6Fe^{2+} + 14H^+ \rightarrow 2Cr^{3+} + 6Fe^{2+} + 7H_2O$

Energetics

1 a) $6C_{(s)} + 6H_{2(g)} + 3O_{2(g)} \rightarrow C_6H_{12}O_{6(s)}$
b) $C_{(s)} \rightarrow C_{(g)}$
c) $H^+_{(aq)} + OH^-_{(aq)} \rightarrow H_2O_{(l)}$ or $NaOH + \frac{1}{2}H_2SO_4 \rightarrow \frac{1}{2}Na_2SO_4 + H_2O$

2 $-85\,KJmol^{-1}$

3 $-2600\,KJmol^{-1}$

Have you improved: Answers

Kinetics

1 a) i) Increase.
ii) Acid particles move more quickly meaning there will be more collisions. Also, because the acid particles have more energy more of them will possess the activation energy and so there will be more successful collisions.
b) i) Increase.
ii) More surface area on the calcium carbonate available for a reaction to occur.

2 a) As diagram on page 39.
b) The total number of particles.
c) At a higher temperature the area under the curve (and so the number of particles) above the activation energy is greater.
d) A catalyst provides an alternative pathway for the reaction of lower activation energy.

3 Thermodynamically speaking, it would be expected that methanol would spontaneously oxidise to methanal (because methanal is lower in energy than methanol). However, because the reaction does not occur spontaneously it means that the reactants are kinetically stable, i.e. the reaction has a high activation energy.

Equilibria

1 They are constant (because the reaction is at equilibrium).

2 a) orange
b) Increasing the acid concentration causes the equilibrium to move to the right-hand side.
c) The equilibrium will move to the right-hand side. Decreasing the temperature favours the exothermic pathway.

3 a) decrease
b) Although a lower temperature is favoured when considering thermodynamics, a very low temperature would mean that the rate of reaction would be too slow.

Groups 1 and 2

1 a) Cs^+ is larger than Li^+, therefore has lower charge density and weaker metallic bonds so less energy required to melt.
b) Metals form ions, Cs larger than Li, therefore ionisation energy lower, therefore reaction faster.
c) $Mg(s) + H_2O(g) \rightarrow MgO(s) + H_2(g)$, heat or steam, MgO insoluble so hydroxide not formed.

2 a) $2Ca + O_2 \rightarrow 2CaO$, $CaO + 2HNO_3 \rightarrow Ca(NO_3)_2 + H_2O$, base.
b) Pt wire dipped into solutions and placed into flame, Ca gives brick red colour, Ba apple green colour. Addition of H_2SO_4 produces thick white ppt with Ba *only* as $BaSO_4$ is very insoluble. Ca^{2+} solution remains clear and colourless OR slightly cloudy.
c) Pass CO_2 gas through $Ca(OH)_2$ solution to produce a white ppt of $CaCO_3$, $CO_2 + Ca(OH)_2 \rightarrow CaCO_3 + H_2O$, in excess ppt dissolves $CO_2 + CaCO_3 + H_2O \rightarrow Ca(HCO_3)_2$.

3 a) Ca^{2+} has a higher charge (density) than K^+ and so polarises the nitrate more strongly and so nitrate is less stable. $Ca(NO_3)_2 \rightarrow CaO + 2NO_2 + \frac{1}{2}O_2$
b) Limestone *or* marble *or* calcite reacts with acids to form salts, e.g. $CaCO_3 + H_2SO_4 \rightarrow CaSO_4 + CO_2 + H_2O$ (OR with HNO_3)
c) Be^{2+} is very small, has a very high charge density and is very polarising, distorts Cl^- anion leading to covalent bonding.

Group 7

1 a) Atomic radius of I larger than F so greater shielding, therefore has stronger van der Waals forces so more energy required to melt.
b) Cl atom smaller with less shielding therefore greater nuclear attraction for electrons. Cl_2 stronger oxidant than Br_2 and so used to extract Br^- ions from sea water. $NaBr(aq) + Cl_2(g) \rightarrow NaCl(aq) + Br_2(aq)$
c) $2NaOH + Cl_2 \rightarrow NaCl + NaClO + H_2O$, disproportionation because Cl goes from 0 to -1 *and* $+1$. Cl_2 used in antiseptics, making HCl and sterilising water, ClO^- used for bleach.

2 a) $NaCl + H_2SO_4 \rightarrow NaHSO_4 + HCl$, misty fumes are HCl which turns damp blue litmus paper red.

b) $NaBr + H_2SO_4 \rightarrow NaHSO_4 + HBr$ then $2HBr + H_2SO_4 \rightarrow SO_2 + Br_2 + 2H_2O$, Br^- larger than Cl^- and has less nuclear attraction for outer electrons and so is strong enough to reduce H_2SO_4.

3 a) Addition of $AgNO_3(aq)$, Cl^- produce white ppt of AgCl soluble in dil. $NH_3(aq)$, Br^- forms cream ppt of AgBr insoluble in dil. $NH_3(aq)$.

b) Add excess KI, I^- oxidised to I_2 and can be titrated with $Na_2S_2O_3$ (from burette)

c) $HBr + H_2O \rightarrow H_3O^+ + Br^-$, fully dissociated, HF only weakly dissociated due to strong H–F bond.

Industrial Inorganic Chemistry

1 a) Carbon burns exothermically to provide heat, $2C + O_2 \rightarrow 2CO_2$, CO, reduces iron oxide to produce metal $Fe_2O_3 + 3CO \rightarrow 2Fe + 3CO_2$.

b) Fe is abundant in high quality ores, cheap to extract, strong and durable and so suitable for construction.

c) Rusts in atmospheric conditions, can be alloyed to form resistant steels.

d) $TiCl_4 + 4Na \rightarrow Ti + 4NaCl$, Na or Mg, inert atmosphere e.g. Ar, formation of carbides reduces desirable properties of the metal.

2 a) Bauxite dissolved in NaOH, $Al_2O_3 + 2OH^- + 3H_2O \rightarrow 2Al(OH)_4^-$, $SiO_2 + 2OH^- \rightarrow SiO_3^{2-} + H_2O$, basic impurities filtered off, then CO_2 blown through solution, $2Al(OH)_4^- + CO_2 \rightarrow 2Al(OH)_3 + CO_3^{2-} + H_2O$.
$Al(OH)_3$ ppt filtered off and heated to form Al_2O_3.

b) Molten cryolite, 850 °C, cathode: $Al^{3+} + 3e^- \rightarrow Al$;
anode: $2O_2^- \rightarrow O_2 + 4e^-$

c) Al forms strong bonds with O and cannot be reduced chemically at a viable temperature, plants sited near cheap sources of electricity e.g. HEP.

3 a) 3–500 °C and 2–1000 atm pressure, temperature a compromise between high rate (at high temperatures) and high yield (low temperatures), high pressures give good yield but are expensive.

b) $N_2 + 3H_2 \rightarrow 2NH_3$, un-reacted N_2 and H_2 recycled.

c) Fe catalyst increases rate of reaction. Impurities (e.g. S) must be removed from feedstock to prevent poisoning

d) Catalytic oxidation of NH_3 to NO. HNO_3 used for making explosives and fertilisers.

Have you improved: Answers

Basic Organic Chemistry and Alkanes

1 a) Same general formula, chemical properties, backbone structure.

b) Hydrocarbon chain is non-polar, so as it increases, less of the molecule can interact with H_2O.

c) Melting/boiling point. As molecule becomes larger intermolecular forces increase so more energy required to separate.

d) $CH_3CH_2CH(OH)CH_3$ = butan-2-ol, $CH_3CH_2C(OH)(CH_3)_2$ = 2-methylpropan-2-ol
$(CH_3)_3$ C-OH = 2,2 dimethyl propanol[8] (any 2)

2 a) Do not contain functional group only strong C–C and C–H bonds. Alkanes combust exothermically therefore releasing energy.

b) $C_7H_{16} + 11O_2 \rightarrow 7CO_2 + 8H_2O$, incomplete combustion forms CO which is toxic.

c) NO can lead to photochemical smog and acid rain. Catalytic converter reduces to N_2 which is harmless $2NO + 2CO \rightarrow N_2 + 2CO_2$.

3 a) Initiation, UV breaks Cl–Cl bond to produce reactive radicals Cl·.

b) $CH_4 + Cl^{\bullet} \rightarrow CH_3^{\bullet} + HCl$, $CH_3^{\bullet} + Cl_2 \rightarrow CH_3Cl + Cl^{\bullet}$

c) Further substitution during propagation, CH_2Cl_2 Mr = 85, $CH_3Cl + Cl_2 \rightarrow CH_2Cl\cdot +$ HCl then $CH_2Cl\cdot + Cl_2 \rightarrow CH_2Cl_2 + Cl\cdot$, removed by fractional distillation.

Alkenes

1 a) Overlap of p-orbitals on C-atoms to produce a region of electron density above and below the C–C sigma bond.

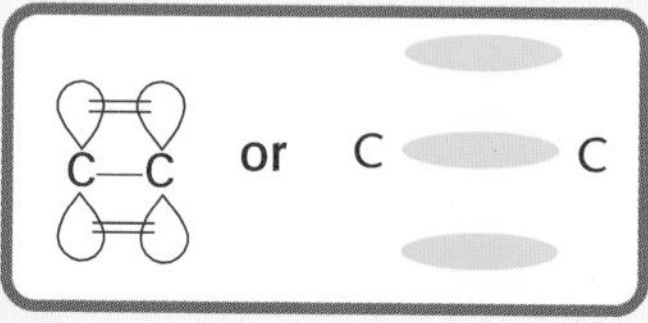

b) Electron density of double bond attracts electrophiles and the weak pi-bond easily breaks, donating an electron pair to the electrophile.

c) Double bond cannot rotate without breaking pi-bond so groups are locked into position.

CH3, H, C=C, H, CH3 — cis
CH3, CH3, C=C, H, H — trans

2 a) $C_4H_8 + HBr \rightarrow C_2H_5CH(Br)CH_3$ 2-bromobutane, HBr gas or conc. solution at RT.

b)

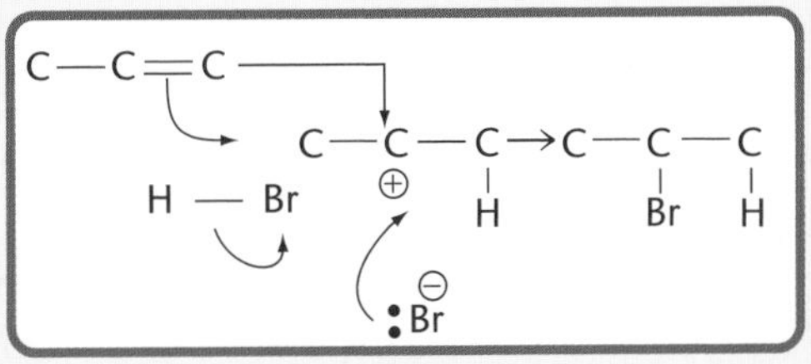

C—C=C → C—C(⊕)—C(H) → C—C(Br)—C(H)

H — Br

$:Br^-$

c) 2-bromopropane is formed via a secondary carbocation intermediate which is more stable than the primary due to extra positive inductive effects from alkyl groups.

3 a) Add a solution of Br_2 in water or CCl_4 to alkene. Red-brown colour disappears.

b) $C_2H_4 + \frac{1}{2}O_2 \rightarrow (CH_2)_2O$, $(CH_2)_2O + H_2O \rightarrow CH_2(OH)CH_2OH$ used as antifreeze in car engines.

c)

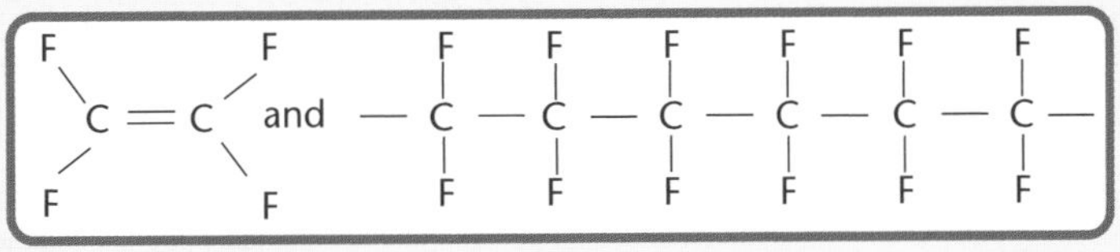

$F_2C{=}CF_2$ and $-CF_2-CF_2-CF_2-CF_2-CF_2-CF_2-$

Free radical reaction, non-stick coatings for kitchenware.

Haloalkanes

1 a) NaOH aqueous, warm, $C_4H_9Br + NaOH \rightarrow C_4H_9OH + NaBr$.

b) OH^- ion is a nucleophile.

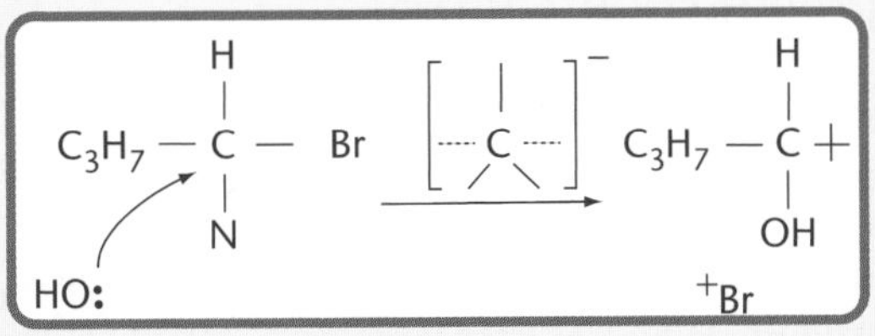

HO: → C_3H_7—C(H)(N)—Br → [transition state]$^-$ → C_3H_7—C(H)(OH) + ^+Br

c) I larger than Br therefor C–I bond weaker, breaks more easily, lowering activation energy therefore faster reaction.

2 a) $C_2H_5Cl + NH_3 \rightarrow C_2H_5NH_2 + HCl$

b) Amines have a lone pair and are also nucleophiles, undergoing further substitution to produce secondary amines, etc.

c) $(C_2H_5)_4N^+Cl^-$, quaternary ammonium salt, used as cationic detergents.

4 a) $2P + 3Cl_2 \rightarrow 2PCl_3$, redox.

b) Simple molecular structure so only weak van der Waals forces are broken; small amount of energy required. No charged, mobile particles therefore cannot conduct electricity.

c) W = $BaCl_2$ X = $CaCl_2$ Y = $Ca^{2+}(aq)$ Z = $BaSO_4(s)$, both chlorides are ionic and dissolve to form hydrated ions, $BaSO_4$ is very insoluble and precipitates, $CaSO_4$ soluble so Ca^{2+} remains in solution. Ca emission spectra shows brick red colour.